Extraction and Fractionation Processes of Functional Components in Food Engineering

Extraction and Fractionation Processes of Functional Components in Food Engineering

Editors

Juliana Maria Leite Nobrega de Moura Bell
Blanca Hernández-Ledesma
Roberta Claro da Silva

MDPI • Basel • Beijing • Wuhan • Barcelona • Belgrade • Manchester • Tokyo • Cluj • Tianjin

Editors

Juliana Maria Leite Nobrega
de Moura Bell
University of California at
Davis
USA

Blanca Hernández-Ledesma
Consejo Superior de
Investigaciones Científicas
(CSIC)
Spain

Roberta Claro da Silva
North Carolina A&T State
University
USA

Editorial Office
MDPI
St. Alban-Anlage 66
4052 Basel, Switzerland

This is a reprint of articles from the Special Issue published online in the open access journal *Processes* (ISSN 2227-9717) (available at: https://www.mdpi.com/journal/processes/special_issues/functional_components_extraction_processes).

For citation purposes, cite each article independently as indicated on the article page online and as indicated below:

LastName, A.A.; LastName, B.B.; LastName, C.C. Article Title. *Journal Name* **Year**, *Volume Number*, Page Range.

ISBN 978-3-0365-5759-5 (Hbk)
ISBN 978-3-0365-5760-1 (PDF)

Contents

About the Editors

Juliana Maria Leite Nobrega de Moura Bell

Juliana Maria Leite Nobrega de Moura Bell, Ph.D; Associate Professor: Dr. de Moura Bell research includes the development and application of environmentally friendly technologies to replace the incumbent technology for extracting and fractionating of major food components such as oil, protein, and carbohydrates. Her goal is to develop structure/function-based processes to produce foods that will improve human health, with the translation of these processes into the industrial realm being the ultimate goal of her work. Specifically, she is interested in bio-processing techniques such as enzyme-assisted aqueous extraction, fermentation, and less harsh techniques like supercritical and subcritical extractions. Her laboratory research interests include: (1) Scaling-up extraction and downstream recovery processes from laboratory to pilot-scale; (2) Determining the effects of processing conditions (extraction, heat treatment, enzymatic modifications and recovery strategies) on the functionality and biological activities of food components, and (3) The conversion of agricultural waste streams/food processing by-products into high added-value compounds.

Blanca Hernández-Ledesma

Blanca Hernández-Ledesma, PhD. Tenured Scientist. Head of the Group Development and Innovation in Alternative Proteins (INNOVAPROT) at CIAL (CSIC-UAM). Her research has been focused on the biological activity of food proteins/peptides aiming at a better understanding of their health implications and the development of novel food ingredients. Author of 90 JCR articles, 37 book chapters, and 3 patents (h-index of 39, WoS). Her results have been presented in 90 international and national conferences. She has supervised 4 Doctoral Thesis (4 are currently ongoing) and 17 Master Thesis. She has participated in 30 international and national Research Projects. She has participated as member of Selection Board for Tenured Scientists, PhD and Master Thesis Dissertation committees, reviewer in international PhD theses, and member of National and International Projects Evaluation Panels. She is member of the Editorial Committees of several books and journals, and reviewer for more than 90 journals.

Roberta Claro da Silva

Roberta Claro da Silva, Ph.D. Assistant Professor: Dr. Silva's research includes lipid science and technology and the physical properties of foods. Her research focuses on developing lipid alternatives to replace saturated and trans fats in food products to improve consumers' health. Dr. Silva's goal is to develop environmentally friendly technologies to replace traditional processes by translating these processes into the industrial realm. Specifically, she is interested in techniques such as oleogelation, nanoemulsification, and sonication. Dr. Silva's team expects to substitute highly saturated fat in food products such as dairy, processed meat, and bakery products without compromising the expectations associated with these conventional food products. Her laboratory research interests include 1. Understand the fat crystallization of structured oleogels and emulsions as a function of time, temperature, and shear using a microstructural approach. 2. Examine both oil-in-water and water-in-oil emulsions, whereby we aim to improve their stability and understand the functional role of various ingredients. 3. Apply the alternative lipids sources in food matrixes and evaluate the impact on the physical and sensorial quality.

Editorial

Introduction to the Special Issue "Extraction and Fractionation Processes of Functional Components in Food Engineering"

Blanca Hernández-Ledesma [1], Roberta Claro da Silva [2] and Juliana Maria Leite Nobrega De Moura Bell [3,*]

1   Department of Bioactivity and Food Analysis, Institute of Food Science Research (CIAL, CSIC-UAM, CEI UAM + CSIC), 28049 Madrid, Spain; b.hernandez@csic.es
2   Family and Consumer Sciences Department, College of Agriculture and Environmental Sciences (CAES), North Carolina A&T State University, Greensboro, NC 27411, USA; rcsilva@ncat.edu
3   Food Science and Technology and Biological and Agricultural Engineering Departments, University of California at Davis, Davis, CA 95616, USA
*   Correspondence: jdemourabell@ucdavis.edu

Citation: Hernández-Ledesma, B.; da Silva, R.C.; De Moura Bell, J.M.L.N. Introduction to the Special Issue "Extraction and Fractionation Processes of Functional Components in Food Engineering". *Processes* **2022**, *10*, 1425. https://doi.org/10.3390/pr10071425

Received: 30 June 2022
Accepted: 10 July 2022
Published: 21 July 2022

Publisher's Note: MDPI stays neutral with regard to jurisdictional claims in published maps and institutional affiliations.

Diet plays an unquestionable role in the growth, development, and maintenance of all body functions. Foods are a source of a multitude of compounds, such as proteins, lipids, carbohydrates, peptides, oligosaccharides, and antioxidants, among others, that go beyond basic nutrition. Harnessing the full potential of the diversity of the nutrients in food becomes essential to enable its use as prophylactic therapy, with the potential to minimize the incidence of several metabolic disorders affecting humans. Food processing is evolving from a traditional approach, with the technical aspects as the main focus, toward a bio-guided strategy, which strives to maintain or improve the original biological and functional properties of food compounds. The development of a holistic approach with cost-competitive yet bio-guided food processing strategies is needed to deliver healthy and nutritious food for everyone.

This Special Issue of Journal Processes includes seven outstanding papers describing novel advances in the development and application of innovative processing strategies, in order to extract, isolate, and modify food compounds to produce ingredients and foods with improved nutritional, functional, and biological properties.

In the review of Franca-Oliveira et al. [1], the existing evidence of suitable, cost-effective, and environmentally friendly technologies to extract high concentrations of valuable proteins from traditional and alternative natural sources is summarized. Novel and eco-sustainable approaches are compared with conventional methods, describing their advantages and current limitations. Moreover, in this article, the combination of these methods with enzymatic hydrolysis is described as a successful strategy to release bioactive peptides at high yield and concentration, which could enable their incorporation into food products and supplements to prevent/treat chronic diseases of high incidence and mortality in our society. One of these eco-friendly extraction technologies is the aqueous extraction process, which relies on the use of upstream mechanical treatments to disrupt the matrix and facilitate the release of intracellular compounds into water. This sustainable approach is used by Dias et al. [2] to concurrently extract oil and protein from almond flour. The proof of concept of this process was demonstrated at a 7 L scale with respect to oil and protein extractability and their distribution amongst the fractions that were generated. Moreover, the impact of enzymatic and chemical demulsification approaches was evaluated on the physicochemical properties and stability of the emulsion proteins and final oil recovery. Aqueous extraction, followed by chemical or enzymatic demulsification, produced a final oil with a similar fatty acid composition to hexane extracted lipids.

In addition to soybean oil, there has been an increasing interest in the use of techniques with a reduced environmental and economic impact to produce plant-based oils such as coconut, flaxseed, and hemp seed as a source of added-value oils. Although expeller pressing and flammable solvent extractions are common methods that are used for oil extraction in the food industry, the low yield for expeller pressing and the environmental impact resulting from the use of organic solvents remain the main concerns limiting the current application of both traditional methodologies, highlighting the importance of alternative green alternatives such as supercritical CO_2 and enzyme-assisted extractions. In the review of Lavenburg et al. [3], the advantages and disadvantages of conventional and novel eco-friendly approaches applied to extract oil from seed by-products are summarized from economic, environmental, and practical perspectives.

Similarly, other food wastes can be a source of bioactive compounds for subsequent recovery and industrial/nutraceutical applications. In the review of Risner et al. [4], the potential of untreated whey permeate as a growth medium for recombinant *Escherichia coli* is described as an affordable approach to microbially produce pinene, a secondary plant metabolite with functional properties as a flavor additive and cognitive health benefits. This process would allow valorizing a dairy by-product by converting it into a valuable co-product, while reducing the detrimental environmental impact that is associated with its disposal. Seeking the development of circular processes, the research of Truong et al. [5] aimed at isolating acid-soluble collagen from snakehead fish (*Channa striata*) by-products, including skin and skin–scale mixtures. The functional and rheological properties of extracted collagen are evaluated, showing that proteins that are obtained from marine sources have the potential to be used as a promising alternative to mammalian collagens.

In the study of Handa et al. [6], the angiotensin-converting enzyme (ACE) inhibitory potential of fermented and non-fermented soy products and isolated 7S and 11S protein fractions is evaluated after their digestion under simulated gastrointestinal conditions. The results of this research provide evidence supporting the important biological role of these products as new functional foods or ingredients to prevent and/or control hypertension and associated diseases.

In addition to extraction methods, in this Special Issue, techniques that are applied for the characterization of the extract compounds are included. The study of Félix-Palomares and Donis-González [7] focuses on the application of Response Surface Methodology (RSM), a multivariate statistic technique, in combination with a Box–Behnken experimental design to optimize and validate the Rancimat operational parameters such as sample weight, temperature, and airflow. The optimized approach allows the identification of standard operational parameters for a more precise, accurate, and efficient determination of walnut oil induction time (resistance of lipids to be oxidized) estimations.

Author Contributions: Writing—original draft preparation, B.H.-L.; writing—review and editing, B.H.-L., R.C.d.S. and J.M.L.N.D.M.B. All authors have read and agreed to the published version of the manuscript.

Funding: This research received no external funding.

Data Availability Statement: Not applicable.

References

1. Franca-Oliveira, G.; Fornari, T.; Hernández-Ledesma, B. A review on the extraction and processing of natural source-derived proteins through eco-innovative approaches. *Processes* **2021**, *9*, 1626. [CrossRef]
2. Dias, F.F.G.; de Almeida, N.M.; de Souza, T.S.P.; Taha, A.Y.; de Moura Bell, J.M.L.N. Characterization and demulsification of the oil-rich emulsion from the aqueous extraction process of almond flour. *Processes* **2020**, *8*, 1228. [CrossRef]
3. Lavenburg, V.M.; Rosentrater, K.A.; Jung, S. Extraction methods of oils and phytochemicals from seeds and their environmental and economic impacts. *Processes* **2021**, *9*, 1839. [CrossRef]
4. Risner, D.; Marco, M.L.; Pace, S.A.; Spang, E.S. The potential production of the bioactive compound pinene using whey permeate. *Processes* **2020**, *8*, 263. [CrossRef]
5. Truong, T.M.T.; Nguyen, V.M.; Tran, T.T.; Le, T.M.T. Characterization of acid-soluble collagen from food Processing by-products of snakehead fish (*Channa striata*). *Processes* **2021**, *9*, 1188. [CrossRef]

6. Handa, C.L.; Zhang, Y.; Kumari, S.; Xu, J.; Ida, E.I.; Chang, S.K.C. Comparative study of angiotensin I-converting enzyme (ACE) inhibition of soy foods as affected by processing methods and protein isolation. *Processes* **2020**, *8*, 978. [CrossRef]
7. Félix-Palomares, L.; Donis-González, I.R. Optimization and validation of Rancimat operational parameters to determine walnut oil oxidative stability. *Processes* **2021**, *9*, 651. [CrossRef]

Review

The Potential Production of the Bioactive Compound Pinene Using Whey Permeate

Derrick Risner, Maria L. Marco, Sara A. Pace and Edward S. Spang *

Department of Food Science and Technology, University of California, Davis, CA 95616, USA;
drisner@ucdavis.edu (D.R.); mmarco@ucdavis.edu (M.L.M.); sspace@ucdavis.edu (S.A.P.)
* Correspondence: esspang@ucdavis.edu; Tel.: +1-530-752-1255

Received: 26 December 2019; Accepted: 19 February 2020; Published: 26 February 2020

Abstract: Pinene is a secondary plant metabolite that has functional properties as a flavor additive as well as potential cognitive health benefits. Although pinene is present in low concentrations in several plants, it is possible to engineer microorganisms to produce pinene. However, feedstock cost is currently limiting the industrial scale-up of microbial pinene production. One potential solution is to leverage waste streams such as whey permeate as an alternative to expensive feedstocks. Whey permeate is a sterile-filtered dairy effluent that contains 4.5% weight/weight lactose, and it must be processed or disposed of due its high biochemical oxygen demand, often at significant cost to the producer. Approximately 180 million m^3 of whey is produced annually in the U.S., and only half of this quantity receives additional processing for the recovery of lactose. Given that organisms such as recombinant *Escherichia coli* grow on untreated whey permeate, there is an opportunity for dairy producers to microbially produce pinene and reduce the biological oxygen demand of whey permeate via microbial lactose consumption. The process would convert a waste stream into a valuable coproduct. This review examines the current approaches for microbial pinene production, and the suitability of whey permeate as a medium for microbial pinene production.

Keywords: terpene; pinene; *Escherichia coli*; whey; whey permeate; biosynthesis; microbial

---

1. Introduction

Terpenoids/isoprenoids are multifunctional organic compounds that contribute to an array of applications and are currently used as solvents, fragrances, natural pesticides, lubricants, flavoring agents, and in nutraceutical/medical applications [1–4]. Terpenoids/isoprenoids are hydrocarbons comprised of five carbon isoprene units and are classified by their number of carbon atoms, namely hemiterpenoids (C_5), monoterpenoids (C_{10}), sesquiterpenoids (C_{15}), diterpenoids (C_{20}), triterpenoids (C_{30}) tetraterpenoids (C_{40}) and polyterpenoids (C45+). Terpenoid precursors are produced in most organisms via the mevalonate (MVA) (Figure 1) and/or the 2-C-methyl-D-erythrithol 4-phosphate/1-deoxy-D-xylose 5 phosphate (MEP/DOXP) pathway (Figure 2) and are synthesized via terpenoid synthases [5]. Pinene is synthesized from isopentenyl pyrophosphate produced from the MVA or MEP/DOXP pathway via geranyl diphosphate synthase and pinene synthase (Figure 3).

Pinene is a monoterpene with two structural isomers, α-pinene and β-pinene (IUPAC names: (2,6,6-trimethylbicyclo[3.1.1]hept-2-ene and 6,6-dimethyl-2-methylenebicyclo[3.1.1]heptane, respectively). Both α-pinene and β-pinene have multiple commercial applications. They are commonly used as flavor compounds to produce herbal or earthy flavors in food [6]. Additional potential commercial applications include uses as a antimicrobial agent, potential use as a jet fuel alternative and functionality as a starting compound for the synthesis of other terpenoids [3,7,8]. This review will examine the potential industrial applications of pinene, the current approaches for microbial pinene production, and the suitability of whey permeate as a medium for microbial pinene production.

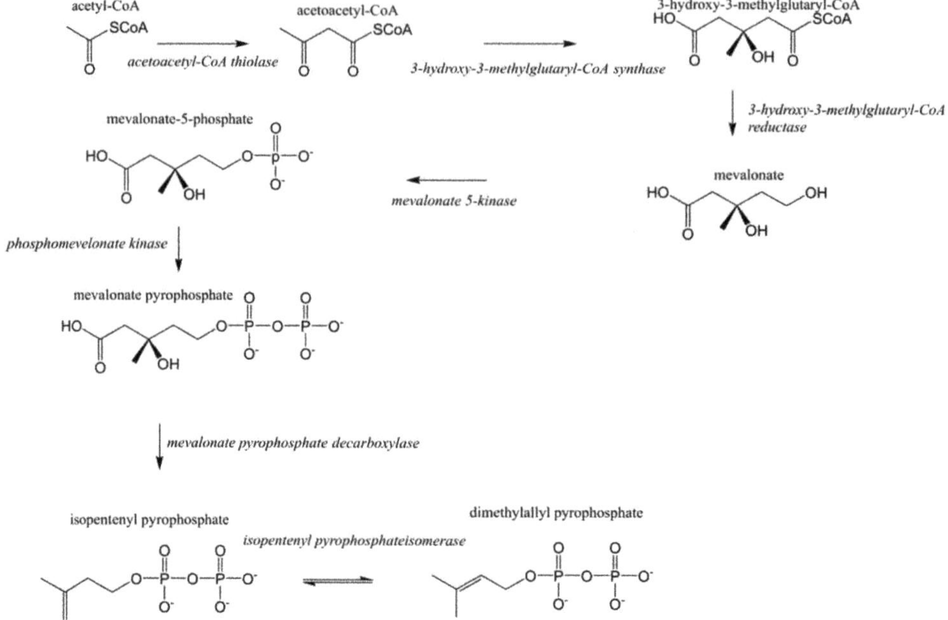

Figure 1. Mevalonate (MVA) pathway for eukaryotes.

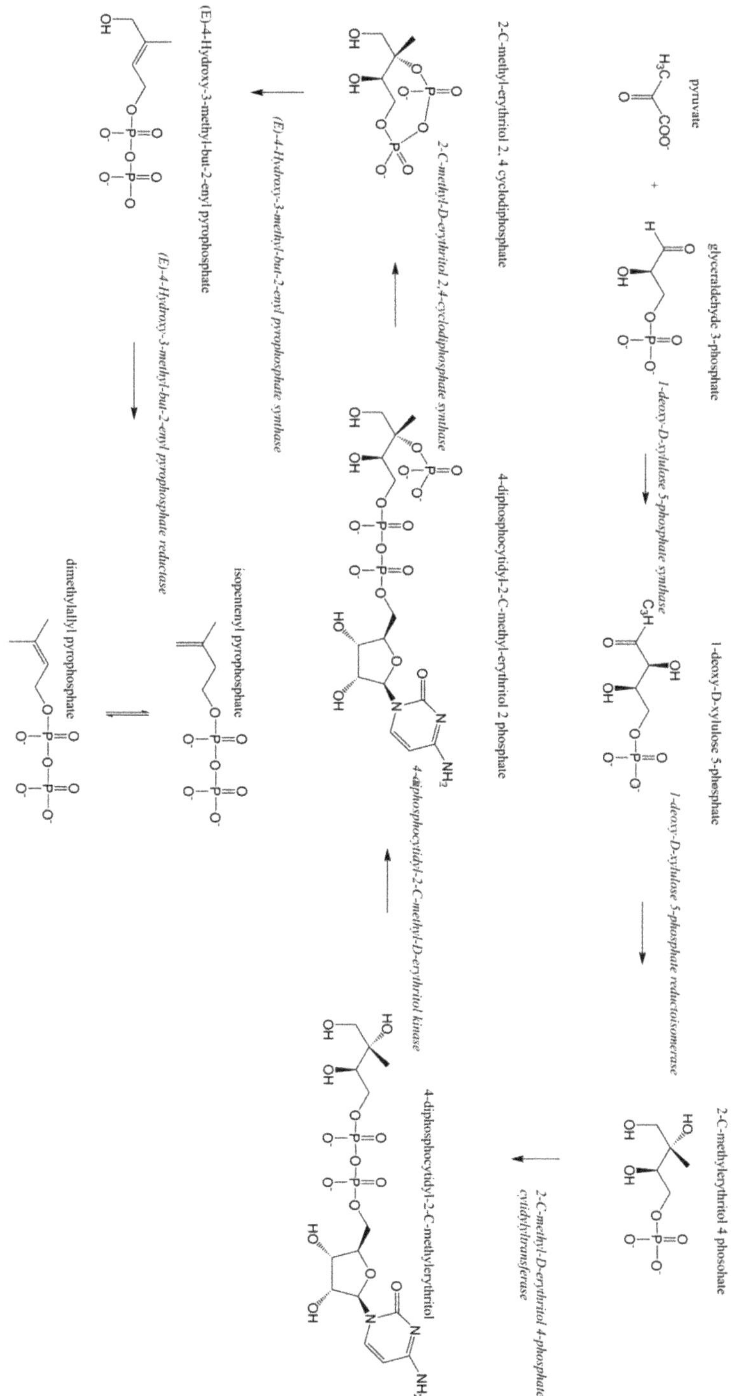

Figure 2. Methylerythritol phosphate (MEP/DOX) pathway.

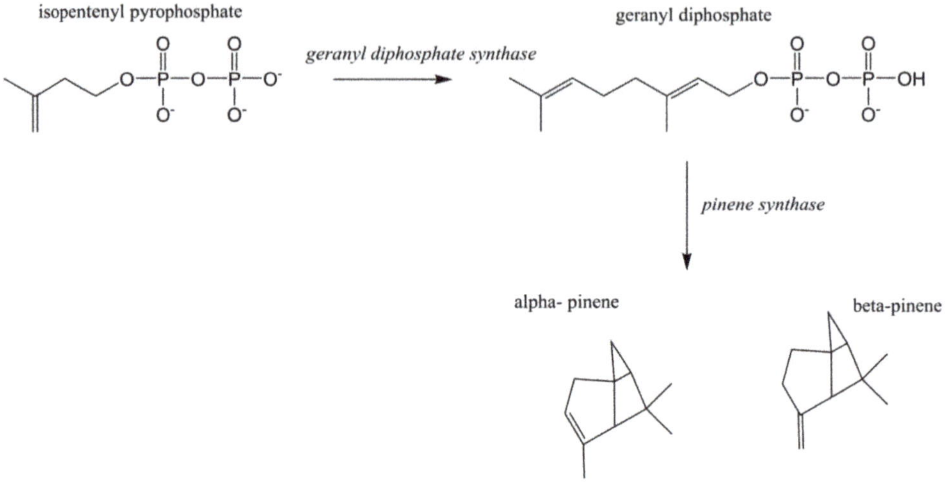

Figure 3. Pinene biosynthesis from isopentenyl pyrophosphate.

2. Applications of Pinene

Pinene is a compound with several current and potential applications in medicine, as a flavor/aroma agent, and as a precursor for fuel production. α- and β-Pinene have several potential applications in medicine. Firstly, it exhibited antimicrobial properties and increased the effectiveness of commercial antibiotics against methicillin resistant *Staphylococcus aureus* [7]. α-pinene also reversibly inhibited acetylcholinesterase, and therefore has potential to treat neurodegenerative diseases such as Alzheimer's disease and Parkinson's disease [1,9]. Additionally, α-Pinene reduced the impact of scopolamine-induced cognitive impairment in mice and 6-hydroxydopamine-induced Parkinson's disease in rats [9,10]. Pinene itself is used as flavoring agent and in cosmetics, but is also used to derive the additional aromatic compounds, verbenone and carvone, which are used in perfumery and as insect repellents [8,11,12]. Collectively, this evidence indicates that pinene has flavoring applications, potential medical value, or could be used as an ingredient in nutraceutical products.

Dimerized pinene also has a volumetric energy similar to that of commercial jet fuel (JP-10) [3]. Although it is not applied commercially as a fuel at this time, increased availability of dimerized pinene could assist airlines in shifting from reliance on fossil fuels towards biofuels. For dimerized pinene to be an economically viable fuel, a sustainable production method must be developed. The combined annual production volume of α and β-pinene is approximately 49 million liters [13]. This is several orders of magnitude less than what would be needed to make a tangible contribution to the jet fuel market, considering that U.S. airline carriers consumed 67 billion liters of jet fuel in 2016 [14].

3. Plant Pinene Biosynthesis and Purification

Pinene is a secondary plant metabolite produced in a wide variety of aromatic plants in small concentrations, however it is produced industrially during the paper making process [15]. Plant-derived isopentenyl diphosphate (IPP) and dimethylallyl diphosphate (DMAPP) are the precursors of pinene and other terpenoids [16]. Geranyl diphosphate (GPP), a monoterpenoid precursor, is formed through the condensation of IPP and DMAPP, which is catalyzed by a terpene synthase (GPPS). GPP is cyclized by pinene synthase (PS) to form α- or β-pinene (Figure 3) [16]. IPP and DMAPP are biosynthesized in plants via the MVA pathway or the MEP pathway [5]. The MVA pathway is initiated by the condensation of two acetyl coenzyme A (acetyl CoA) molecules (Figure 1). The MEP pathway is initiated by the condensation of glyceraldehyde 3-phosohate and pyruvate (Figure 2). Glyceraldehyde 3-phosohate, pyruvate and acetyl CoA are inputs into several metabolic pathways (glycolysis, tricarboxylic acid cycle, gluconeogenesis and the pentose phosphate pathway), thus the MVA and MEP pathways are heavily regulated [5]. Plants can utilize both pathways for terpenoid production; however, pinene and other terpenoids are produced in relatively small quantities in plants from these pathways due to metabolic regulation [3,5].

Currently, in the United States, pinene is produced as a byproduct from the wood pulp production process known as kraft pulping [6,17,18]. Turpentine is collected during the heated digestion process for wood pulp production and pinene can be extracted from turpentine via fractional vacuum distillation [19]. One ton of fresh pine wood yields between 6–18 L of turpentine [20]. Turpentine is composed of a variety of terpenoids and its concentration is variable. α-Pinene is the primary constituent (75%–85%) with variable amounts of β-pinene (up to 3%), camphene (4%–15%), and limonene (5%–15%) [21]. Pinene can also be found in low concentrations in variety of other plants including rosemary (*Rosmarinus officinalis*), sage (*Salvia officinalis*) [22], wild thyme (*Thymus serpyllum*) and a variety of conifers (plants from the class Pinosida) [23]. Additional refining and processing are required for pinene to be used as a flavor ingredient, nutraceutical, or fuel additive.

4. Microbial Pinene Biosynthesis

With only small quantities produced in plants, there is an opportunity to produce pinene and other terpenoids microbially. A common strategy for the microbial production of pinene is the transformation of plasmids containing genes isolated from pinene producing plants [6,8,16,24].

These plasmids encode for the production of the enzymes within the pathways illustrated in Figures 1–3, which enables for the production of the intermediate compounds and subsequently, pinene. The precursors to these pathways are pyruvate and acetyl-CoA, products of the glycolysis and citric acid cycle, which are commonly utilized for cellular energy production. A portion of pyruvate and acetyl-CoA produced during cellular metabolism can be potentially utilized for microbial pinene production. For example, pinene could be produced in a similar manner to the microbial production of β-farnesene, another terpenoid [25]. Currently, recombinant *Saccharomyces cerevisiae* is being used to produce β-farnesene in 200,000 L bioreactors for industrial production; β- farnesene is then used in variety of commercial applications [25]. The following sections discuss the pros and cons of utilizing some promising organisms for the microbial production of pinene and highlights the strategies used for microbial engineering (Table 1).

Table 1. Pros and cons of potential recombinant microorganisms that produce pinene and their use of whey permeate as medium *.

		Recombinant Microorganisms				
		E. coli [6,16,24,26]	S. cerevisiae [24,27–30]	C. glutamicum [31]	R. toruloides [32,33]	K. marxianus [34,35]
Pros	Genetic traceability	+	+			
	Generally recognized as safe		+	+	+	+
	Focus of current research	+				
	Metabolizes lactose	+				+
	Metabolizes galactose	+	+		+	+
	Reported levels of pinene production greater than 1 mg/L	+				
	Used for commercial production of ethanol or other metabolites	+	+	+		+
	Commercial production using whey permeate as medium					+
Cons	Potential extraction difficulties due to outer membrane	−				
	LPS (endotoxin)	−				
	Native MVA pathway with competing enzymes		−			−
	Little to no research related to pinene production		−	−		−

* Table was derived from current literature concerning pinene production from recombinant microorganisms [6,16,24,26–35].

4.1. E. coli

The MVA pathway was expressed in *E. coli* using genes from *Enterococcus faecalis* (*mvaE*, *mvaS*, *mvaE*) and *S. cerevisiae* (*ERG12*, *ERG8*, *ERG19*, *IDI1*) ([6,8,16,26]. Geranyl diphosphate synthase (GPPS) and pinene synthase (PS) were expressed using *GPPS2* and *Pt30* genes derived from *Abies grandis* and *Pinus taeda*, respectively [8]. GPPS catalyzes the conversion of isopentenyl pyrophosphate, also known as isopentenyl diphosphate (IPP) into geranyl diphosphate (GPP), which is the substrate for PS whose final product is pinene [16]. Using the expressed pathways described above, an accumulation of 5.44 mg/L of E. coli produced α-pinene was reported [8].

A 30-fold increase in *E. coli* pinene production was reported from 2013 to 2018 (Figure 4). These advancements were initiated with screening PS and GPPS from three different species of plants in a strain of *E. coli* harboring the MVA pathway [6]. *E. coli* expression of a combination of PS and GPPS from *Abies grandis* resulted in the production of 28 mg/L of pinene [6]. It was hypothesized that pinene production was limited by the toxicity of GPP, an intermediate in the pinene production pathway. To overcome GPP toxicity, a GPPS-PS protein fusion was produced in order to reduce GPPS inhibition [6]. That step resulted in an increased pinene yield of 32 mg/L [6].

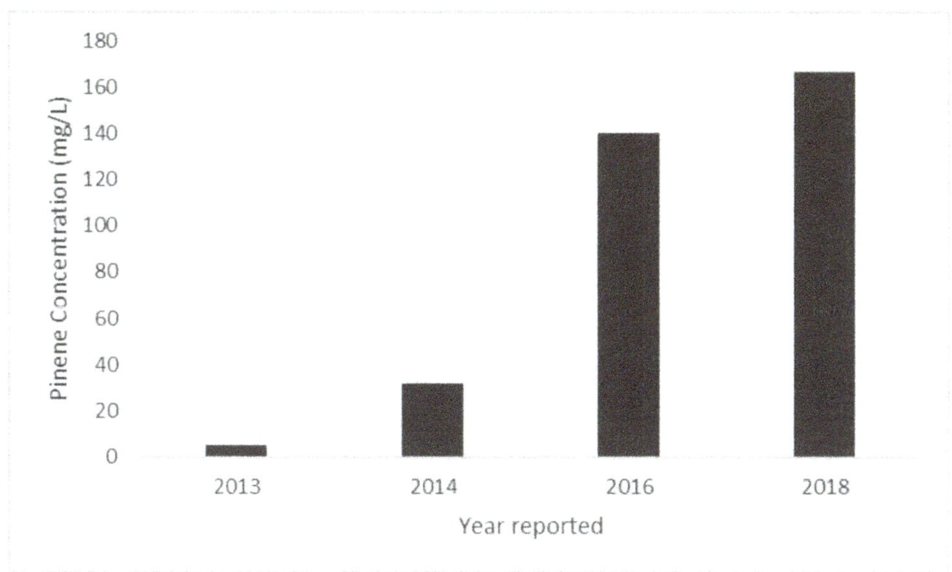

Figure 4. Reported pinene concentration produced by genetically engineered *E. coli* in a single batch laboratory production system. Values obtained from the following reports: 2013—Reported in *Biotechnology for Biofuels* [8], 2014—Reported in *ACS Synthetic Biology* [6], 2016—Reported in *ACS Synthetic Biology* [26], 2018—Reported in *Frontiers in Microbiology* [16], reported standard deviation ± 0.3.

An even higher level of pinene synthesis by *E. coli* (140 mg/L) was subsequently reported, due to the introduction of a laboratory-evolved pinene synthase variant with a higher activity than the parent enzyme native to *Pinus taeda* [26]. Most recently, *E. coli* production of pinene at levels of 166 mg/L was shown using two recombinant *E. coli* strains. One strain expressed the genes for the MVA pathway and the other expressed the genes for GPPS and PS [16]. Each strain was propagated separately and then were combined into a whole cell biocatalyst that was a phosphate buffer overlaid with dodecane. The dodecane accumulated 166 mg/L of pinene after 28 h [16]. This finding indicates that the scale-up of the whole cell biocatalyst system may warrant further research for pinene production.

There are several challenges related to increasing *E. coli*-produced pinene titers. The heterologous enzymes introduced to *E. coli* for pinene production produce intermediates that cannot be utilized by the *E. coli* cell and this metabolic burden can cause plasmid instability [36]. Plasmid recombination can be reduced and pinene titer increased by altering the transcriptional rrnB-T1 terminators on the plasmid to strong terminators that are not homologous [36]. An *E. coli strain* with the genes *recA*, *recE*, *recF*, *recI* and *endA* removed was transformed with a plasmid encoded for pinene production with strong terminators. This *E. coli* strain increased pinene titer by 6.9 fold [36].

The catalytic process of mevalonic kinase and mevalonate diphosphate decarboxylase were identified as the bottlenecks of the MVA pathway in recombinant *E. coli* [36]. A strain of laboratory involved pinene-producing *E. coli* strain was whole genome sequenced, then aligned with the *E. coli* reference strain MG1655 genome sequence, then non-synonymous mutant genes were up and downregulated [37]. Using comparative genomics and transcriptional level analysis, it was shown that mutations and up-regulation of transcription levels of the MEP pathway genes (*dxr, dxs, ispH,* and *ispU*) and some protein membrane genes (*gsiA, nlpA, sufBCDS, opgB, setC, oppF, ypjK, btuB, pitA* and *cusA*) are related to increased pinene tolerance and production in *E. coli* [37]. Overexpression of genetic information processing genes (*cbpA, rpoA, mutS* and *dusB*) and cellular process genes (*tabA* and *flgFGH*) also improved pinene titer [37]. The mutation and transcriptional downregulation of cellular division modulation genes (*ydiJ, yjbQ, prpR, marR, fabR* and *cedA*) was also associated with increased pinene tolerance and production [37].

Although the use of *E. coli* for pinene production has many advantages, including but not limited to its genetic tractability and its current use for the production of other isoprenoids [24], a limitation with this organism is that *E. coli's* outer membrane is comprised of LPS, a potential food safety concern if pinene is destined for human consumption.

4.2. S. cereviseae and K. marxianus

S. cerevisiae is another commonly used microorganism for genetic engineering, and has been successfully used to commercially produce β-farnesene, a sesquiterpene [25]. Monoterpenoid production has also been reported for genetically engineered *S. cerevisiae* in conjunction with other isoprenoids [29,30,38]. *S. cerevisiae* and *Kluyveromyces marxianus* that are generally recognized as safe may be better candidates for pinene production if the genetic engineering challenges can be overcome. *K. marxianus* can metabolize lactose; however, it has the same genetic engineering challenge as *S. cerevisiae*; the presence of a native competing MVA pathway.

S. cerevisiae is an attractive organism to use for monoterpenoid production due it's genetic traceability and being GRAS. The MVA pathway is native to *S. cerevisiae*, which initially seems promising; however, it is highly regulated and critical for cell survival. Monoterpene synthesis is more difficult to engineer into recombinant *S. cerevisiae* due to the presence of the farnesyl pyrophosphate synthase (Erg20p), an enzyme which competes for GPP, the precursor to pinene and other monoterpenoids [30,39]. Farnesyl pyrophosphate is an intermediate in the ergosterol production pathway and ergosterol plays a critical role in *S. cerevisiae* cell membrane fluidity, permeability, and structure [40]. Deletion of ERG20, the gene responsible for Erg20p production, is lethal [30]. It has been reported that direct downregulation of ERG20 did not increase monoterpene production, and instead may have decreased production [40]. Evidence also indicates that Erg20p has greater activity for GPP than PS, another factor which could limit pinene production [41,42].

Several strategies have been employed to overcome the necessity of ergosterol production and the enzymatic competition of the synthases for *S. cerevisiae* monoterpenoid production. N-degron protein degradation was employed to regulate Erg20p/ERG20; this yielded reported titers of 76 mg/L of limonene [39]. It was also shown that fusion of a terpene synthase with Erg20p increased *S. cerevisiae* monoterpene titers by 69% [41]. Another strategy is to replace the native ERG20 promoter with a heterologous promoter [42]. The native ERG20 promoter was replaced with a glucose sensing promoter (HXT1) in a geraniol producing strain of *S. cerevisiae*. This resulted in 897 mg/L of geraniol being produced when the organisms were fed pure ethanol. It was further reported that a titer of 1.69 g/L was achieved via gene deletion (*OYE2*) and engineering for complemented LEU2 auxotrophy [42]. These strategies could be employed to produce pinene using *S. cerevisiae* as the host organism. Further, it may be advantageous to use a microorganism for which the MVA pathway is not critical for cell survival, or a strain of *S. cerevisiae* with lower Erg20p activity.

4.3. Corynebacterium glutamicum and Rhodosporidium torulides

C. glutamicum is used for microbial amino acid production on a commercial scale GRAS status and, in one study, was used to produce pinene [31]. Pinene synthesis was shown using plasmid-encoded genes for deoxy-D-xylose-5-phosphate synthase, isopentenyl diphosphate isomerase, two geranyl diphosphate synthases, and pinene synthase. The native MEP pathway genes responsible for deoxy-D-xylose-5-phosphate synthase and isopentenyl diphosphate isomerase production (*dxs* and *idi* genes) were also overexpressed. The maximum pinene titer reported was 176 µg/L, a quantity that is several orders of magnitude lower than the pinene production reported for *E. coli* [31]. A potential solution to the lower pinene yields is the introduction of a heterologous MVA pathway into *C. glutamicum*. The use of a heterologous MVA pathway has led to increased pinene titers in *E. coli* [8,16,37]. Introduction of the MVA pathway could increase the ability of a strain of *C. glutamicum* to produce pinene.

Monoterpene production using a strain of carotenogenic yeast, *Rhodosporidium torulides*, was also reported [33]. Recombinant *R. torulides* harboring genes to produce nine different monoterpenes produced very little pinene according to a dodecane overlay analyzed using GC-MS; however, it was detected using solid-phase microextraction. It was determined that the synthesis of GPP, which is an intermediate in ergosterol production, was found to be a limiting factor [33]. Strategies used for *S. cerevisiae* to increase monoterpene production, such as using N-degron protein degradation to regulate Erg20p/ERG20 or replacement of the ERG20 promoter with a heterologous promoter, could be employed.

5. Possible Challenges for Large-Scale Synthesis of Pinene by Microorganisms

Besides increasing the quantities of pinene produced by microorganisms, several other genetic and physiological challenges must be overcome for the microbial production of pinene or other monoterpenoids to be economically feasible. Cell toxicity to pinene must be overcome for pinene production to achieve commercial viability. This might be achieved by strategies such as the serial culturing of cells with increasing pinene concentrations. Serial culturing of *E. coli* improved pinene tolerance and increased pinene titer 31% compared with the starting strain and this may be an applicable strategy for other microorganisms [16]. Overexpression of efflux pumps has been shown to increase tolerance to harmful secondary metabolites. The over expression of efflux pumps which remove pinene from the cell such as *E. coli* acrAB and *Pseudomonas putida* KT2440 TtgB could be employed in *E. coli* or other microorganisms [16]. The over expression of *E. coli* acrAB and *Pseudomonas putida* KT2440 TtgB efflux pumps increased the pinene titer from 7.3 ± 0.2 mg/L to 9.1 mg/L, a 25% increase in E. coli [16].

It will be important for the pinene-producing microorganisms to withstand the fluctuating hydrostatic pressures of industrial bioreactors [25]. Standard industrial bioreactors are reported to contain up to 200,000 L [25,43]. Thermotolerance is also important because removal of heat produced from cellular metabolism is a challenge as bioreactor size is increased [43]. Heat accumulates due to the volume of fluid being much greater than the heat transfer surface area [43]. The addition of oxygen to medium could potentially cause facultative anaerobes to respire which releases more energy (74.4 vs. 2808 kJ per $C_6H_{12}O_6$ molecule metabolized) [44] into the system, which must subsequently be removed. The microorganisms would also ideally grow on an inexpensive and readily available medium.

Recent studies focused on the production of pinene by *E. coli* in a single-batch system [6,8,16,37]. Alternatively, a fed-batch system can be used. A titer of 970 mg/L of *E. coli*-produced pinene was reported in fed-batch fermentation conditions, whereas the same strain in the single-batch system produced 5.44 mg/L [8]. This indicates that a fed-batch system may be more commercially viable on an industrial scale. This yielded a conversion efficiency of 2.61% by mass of pinene to mass of glucose for the fed batch system [8]. The production of harmful metabolites may reduce microbial pinene production efficiency. An example of this was the production acetic acid, which is a harmful metabolic by-product of the *E. coli* in a single-batch system [8]. This was found to be a non-limiting factor in a fed-batch system using the same strain of *E. coli* in a fed-batch system maintained at 0.5 g/L of glucose [8]. These highly controlled conditions are the likely reason for the increase in pinene titer when compared with the single-batch system. These findings indicate that more research is required to understand if increases in pinene concentration in a single-batch bioreactor will translate to increased pinene production in a fed-batch bioreactor or chemostat.

Lastly, increased titers of *E. coli* produced pinene was reported using a laboratory-scale whole cell biocatalyst system [16,37]. This system involved growing strains independently, then combining them into an agitated bioreactor. The biocatalyst system has produced a titer 2.6-fold higher than a fermentation system (166 vs. 64 mg/L) [16]. This indicates that the scale-up of a whole cell biocatalyst system for pinene production should be explored.

The differences in titer in the fed-batch bioreactor and the single-batch system indicate that environmental factors other than genetics influence microbial pinene production yields such as pH, temperature, solute or ion concentration, etc. Pinene producing microorganisms would ideally grow on an inexpensive and readily available medium. Given that feedstock is a major expense for biomanufacturers, the use of an inexpensive feedstock could potentially impact the economic feasibility of microbial pinene production.

6. Whey as a Feedstock for Microbial Pinene Production

A recent techno-economic assessment of microbial terpenoid production indicated that 80%–95% of the costs can be attributed to the cost of the feedstock [13]. Cane syrup is often utilized as the industrial feedstock for *S. cerevisiae* β-farnesene synthesis [25,45], but its cost is a current limitation to commercializing the microbial production of pinene due to lower titers [13]. Thus, the identification of an alternative low-cost feedstock for microbial pinene production could significantly improve the cost-effectiveness.

Whey permeate may be an economically viable feedstock for microbial pinene production. Whey and whey permeate have been used as a microbial medium to industrially produce ethanol since 1978, indicating that it may be an economically viable microbial medium [46]. It is produced in large quantities during cheese production, undergoes a filtration process for protein recovery, and is considered a waste stream in many production facilities. The ultrafiltration and diafiltration of whey produce a sterile permeate composed of water, lactose, non-protein nitrogen and minerals. The sterile filtration of whey and the composition of the permeate indicate that it may be a viable microbial medium. Table 2 illustrates the approximate composition of whey and whey permeate [47]. For most microbes that can utilize lactose, it is hydrolyzed intracellularly into its constituents, glucose and galactose. Glucose can be utilized for cellular energy production via glycolysis and galactose can be converted to glucose-6-phosphate, an intermediate in glycolysis. The product of glycolysis is pyruvate, which can be converted to acetyl-CoA. Both of these compounds serve as the precursors to the MEP and MVA pathways, respectively. Trace amounts of oligosaccharides, peptides and (β-casein, αs1-casein, GLCM1, PIGR, SAA, κ-casein and αs2-casein) and enzymes (plasmin, cathepsin and elastase) have been identified with whey permeate [48,49].

The majority of value-added processing of whey permeate currently focuses on recovering lactose. However, lactose recovery is an energetically expensive process and requires significant capital investment. Lactose recovery from whey permeate is a multi-stage process that requires an evaporator, crystallization tanks, decanter centrifuges, fluid-bed drier and hammer mill [47]. As such, lactose recovery is not a viable economic option for many creameries (especially small to medium-size creameries).

Table 2. Approximate composition of whey and whey permeate.

	Protein (% *w/w*)	Lactose (% *w/w*)	Non-Nitrogen Protein (% *w/w*)	Ash (% *w/w*)	Fat (% *w/w*)
Whey	0.6	4.5	0.20	0.50	0.03
Whey Permeate	0.01	4.5	0.20	0.50	n.d.[a]

Table derived from the Tetra Pak's Dairy Processing Handbook [47]. [a] n.d. = not detectable.

If not used for other purposes, whey permeate is a costly waste stream in the dairy industry, and in some situations can represent an economic limitation to creamery expansion. The biological oxygen demand (BOD) (approximately 30–50 g/L) of whey permeate is the main driver of the high cost of disposal. Importantly, microbial lactose consumption can lower the BOD up to 75% [50], and thus, microbial treatment of whey permeate represents the potential combined benefit of delivering value-added product while simultaneously reducing treatment and disposal costs of whey permeate.

Whey permeate is currently used as a fermentation medium for the production of single cell protein, lactic acid, vitamin B12 and ethanol. More specifically, recombinant *E. coli* strains have produced ethanol using untreated whey permeate as a fermentation medium (Table 1) [51]. This indicates that whey permeate may be a suitable medium for other recombinant *E. coli* strains or other recombinant microorganisms. In fact, multiple studies have demonstrated the ability of *E. coli* to metabolize the lactose within whey permeate [6,8,16,26,52]. It was reported in 2017 that an *E. coli* strain efficiently grew and fermented untreated whey permeate without nutritional supplementation in a pH-controlled bioreactor [51]. Recent efforts also indicate that recombinant homofermentative *E. coli* strains expressing the *Vitreoscilla* hemoglobin successfully produce ethanol using rehydrated whey powder and autoclaved cheese whey as a medium [53,54]. This indicates that recombinant *E. coli* used to produce pinene may be able grow on whey permeate.

S. cerevisiae, the organism traditionally used for ethanol production, cannot metabolize lactose. Native *S. cerevisiae* does not produce lactose permease to facilitate the transmembrane transport of lactose, nor does it produce β-galactosidase to hydrolyze lactose to D-glucose and D-galactose [55,56]. The addition of β-galactosidase to whey permeate can make whey permeate a suitable medium for *S. cerevisiae* [57]. Alternatively, engineered *S. cerevisiae* strains can metabolize lactose directly [55,56,58]. This indicates that a recombinant strain of *S. cerevisiae* may be utilized to produce pinene from whey permeate.

C. glutamicum lacks the necessary genes to metabolize lactose or galactose, but similar to the case with *S. cerevisiae*, recombinant strains of *C. glutamicum* have been produced that can metabolize lactose and galactose directly [59]. The inability of *C. glutamicum* to metabolize galactose indicates that the addition of exogenous β-galactosidase to whey permeate would not be an effective treatment for the medium. The additional metabolic engineering requirement for *C. glutamicum* may make it an inappropriate candidate for microbial pinene production using whey permeate.

R. torulides can metabolize glucose and galactose; however, it grows poorly using whey permeate as a medium [60]. Treatment of whey permeate with exogenous β-galactosidase would likely make the whey permeate a more suitable medium for *R. torulides* as it can metabolize galactose.

The production of ethanol using whey permeate as a feedstock has been accomplished on a commercial scale using *K. marxianus* as the fermentation organism. *K. marxianus* is GRAS and can metabolize lactose [46]. *K. marxianus* could be a suitable recombinant microorganism for the microbial production of pinene using whey permeate as a medium (Table 1). *K. marxianus* is a fast growing microorganism and has been observed to grow at temperatures greater than 50 °C [61–63]. *K. marxianus* has not traditionally been used in genetic engineering; however, *K. marxianus'* genome has been sequenced and tools have been developed for the genetic engineering of *K. marxianus* NBRC1777 [35]. A plasmid-based CRISPR-Cas9 was adapted for the genome editing of *K. marxianus* Km17 [63]. This enabled the engineering of stable heterothallic haploid strains of *K. marxianus* [63]. Recombinant *K. marxianus* KM-L9-20 has been cultured in an optimized whey medium to overproduce lactase [34].

This indicates that whey permeate may be a suitable medium for *K. marxianus* genetically engineered to produce pinene.

7. Whey Standardization and Pinene Recovery

To scale-up pinene production utilizing dairy waste streams, the composition of whey and whey permeate feedstocks would need to be standardized. Evaporators, already commonly used for lactose recovery, could be used to standardize the lactose concentration of the permeate and/or increase the lactose concentration to a desired level. The evaporators could also be used to make a concentrate, which could then be used in a fed-batch bioreactor system. This could make whey permeate a more consistent and desirable microbial medium for pinene production.

Due to the fact that the commercial production of pinene utilizes condensers to collect the volatile compound from heated wood pulp digestors [20,64], the recovery method for pinene from whey would need to employ a different technology. Liquid-liquid extraction is the common method for recovery of microbially produced non-polar compounds [65]. Disruption of the cell wall may increase pinene yields, and due to pinene being a small non-polar compound, counter-current supercritical CO_2 extraction could be used as an alternative recovery process [65,66].

Currently, the highest reported yield for *E. coli* pinene production is 2.61% g pinene/g glucose [8]. If the 10% of annual lactose produced from cheese production (860,000 metric tons) can be utilized in a similar manner as glucose, this would equate to 22,446 metric tons of pinene that could be produced a year. This represents approximately half of the current market for pinene [13]. The 2018 market size for α- and β -pinene was approximately $56 and $102 million/year, respectively [13]. If the challenges of monoterpenoid production can be overcome and microbial production yields are similar to those of the microbially produced sesquiterpene, β-farnesene (17.3% g β-farnesene/g glucose) then this could equate to the production of 148,780 metric tons of pinene [25]. This level of production could lend itself to production as a fuel additive precursor like β-farnesene. This highlights the potential of using whey permeate, a sterile-filtered cheese production by-product, as a microbial medium for recombinant microorganisms to produce pinene. Whey and whey permeate are considered waste streams at many cheese production facilities. The production of pinene using a whey permeate as a medium potentially decreases the BOD of whey, due to the consumption of lactose, and produces a valuable compound that has several different commercial applications. This process could potentially be applied to other microbially produced terpenoids which have health, perfumery, and flavoring applications.

8. Conclusions

Whey permeate, a dairy processing effluent, has the potential to become a feedstock to produce high-value bioactive compounds such as pinene. As the capacity for microbial synthesis of high-value bioactive compounds progresses, the need for a sterile growth medium will become increasingly relevant to biotechnology companies. This opportunity coincides with the need of cheese production facilities to process whey permeate, to reduce its BOD. Such synergy could lead to an opportunity to create a value-added product—pinene—from what is currently a waste stream in many cheese production facilities. Although several challenges must be overcome for the microbial production of pinene or other terpenoids to be economically feasible, they could be overcome by improving genetic methods for non-model organisms that also have GRAS status (e.g., *K. marxianus*). Research related to the scale-up of a co-modular biocatalyst system and a pinene separation system is also needed. An economic evaluation, such as a techno-economic assessment, should be conducted to understand the economic feasibility of this proposed process. An environmental evaluation, such as a life cycle assessment comparing the current processing methods of whey permeate with the proposed process, should be conducted to understand the environmental implications of embracing this emerging technology. This review calls for further research into the use of whey permeate as a microbial medium for recombinant microorganisms that produce plant secondary metabolites.

Author Contributions: Authors contributed in the following roles: conceptualization, D.R., M.L.M., S.A.P. investigation, D.R. writing, original draft preparation, D.R. writing, review and editing, D.R., M.L.M., S.A.P. and E.S.S. supervision, E.S.S. and M.L.M. project administration, E.S.S. funding acquisition, E.S.S. All authors have read and agreed to the published version of the manuscript.

Funding: This research received no external funding.

Conflicts of Interest: The authors declare no conflict of interest.

References

1. Miyazawa, M.; Yamafuji, C. Inhibition of Acetylcholinesterase Activity by Bicyclic Monoterpenoids. *J. Agric. Food Chem.* **2005**, *53*, 1765–1768. [CrossRef]

2. Miller, J.A.; Thompson, P.A.; Hakim, I.A.; Chow, H.-H.S.; Thomson, C.A. d-Limonene: A Bioactive Food Component from Citrus and Evidence for a Potential Role in Breast Cancer Prevention and Treatment. *Oncol. Rev.* **2011**, *5*, 31–42. [CrossRef]

3. George, K.W.; Alonso-Gutierrez, J.; Keasling, J.D.; Lee, T.S. *Isoprenoid Drugs, Biofuels, and Chemicals—Artemisinin, Farnesene, and Beyond*; Springer: Cham, Swizerland, 2015; pp. 355–389.

4. Clomburg, J.M.; Crumbley, A.M.; Gonzalez, R. Industrial Biomanufacturing: The Future of Chemical production. *Science (80-.)* **2017**, *355*, aag0804. [CrossRef]

5. Chang, M.C.Y.; Keasling, J.D. Production of Isoprenoid Pharmaceuticals by Engineered Microbes. *Nat. Chem. Biol.* **2006**, *2*, 674–681. [CrossRef]

6. Sarria, S.; Wong, B.; Martín, H.G.; Keasling, J.D.; Peralta-Yahya, P. Microbial Synthesis of Pinene. *ACS Synth. Biol.* **2014**, *3*, 466–475. [CrossRef]

7. Da Silva, A.C.R.; Lopes, P.M.; de Azevedo, M.M.B.; Costa, D.C.M.; Alviano, C.S.; Alviano, D.S. Biological Activities of a-Pinene and β-Pinene Enantiomers. *Molecules* **2012**, *17*, 6305–6316. [CrossRef] [PubMed]

8. Yang, J.; Nie, Q.; Ren, M.; Feng, H.; Jiang, X.; Zheng, Y.; Liu, M.; Zhang, H.; Xian, M. Metabolic Engineering of *Escherichia coli* for the Biosynthesis of Alpha-Pinene. *Biotechnol. Biofuels* **2013**, *6*, 60. [CrossRef] [PubMed]

9. Goudarzi, S.; Rafieirad, M. Evaluating the Effect of α-Pinene on Motor Activity, Avoidance Memory and Lipid Peroxidation in Animal Model of Parkinson Disease in Adult Male Rats. *Res. J. Pharmacogn.* **2017**, *4*, 53–63.

10. Lee, G.-Y.; Lee, C.; Park, G.H.; Jang, J.-H. Amelioration of Scopolamine-Induced Learning and Memory Impairment by α -Pinene in C57BL/6 Mice. *Evidence-Based Complement. Altern. Med.* **2017**, *2017*, 1–9. [CrossRef] [PubMed]

11. Chen, W.; Vermaak, I.; Viljoen, A. Camphor—A Fumigant during the Black Death and a Coveted Fragrant Wood in Ancient Egypt and Babylon—A Review. *Molecules* **2013**, *18*, 5434–5454. [CrossRef] [PubMed]

12. World Health Organization. *Evaluations of the Joint FAO/WHO Expert Committee on Food Additives-Verbenone*; World Health Organization: Geneva, Switzerland, 2019; p. 1.

13. Wu, W.; Maravelias, C.T. Synthesis and Techno-Economic Assessment of Microbial-Based Processes for Terpenes Production. *Biotechnol. Biofuels* **2018**, *11*, 294. [CrossRef] [PubMed]

14. Schipper, M. *As U.S. Airlines Carry More Passengers, Jet Fuel Use Remains Well Below its Previous Peak—Today in Energy*; U.S. Energy Information Administration (EIA) United States Department of Energy: Washington, DC, USA, 2017.

15. Al-Hamimi, S.; Abellan Mayoral, A.; Cunico, L.P.; Turner, C. Carbon Dioxide Expanded Ethanol Extraction: Solubility and Extraction Kinetics of α-Pinene and cis-Verbenol. *Anal. Chem.* **2016**, *88*, 4336–4345. [CrossRef] [PubMed]

16. Niu, F.-X.; He, X.; Wu, Y.-Q.; Liu, J.-Z. Enhancing Production of Pinene in *Escherichia coli* by Using a Combination of Tolerance, Evolution, and Modular Co-culture Engineering. *Front. Microbiol.* **2018**, *9*, 1623. [CrossRef] [PubMed]

17. Behr, A.; Johnen, L. Myrcene as a Natural Base Chemical in Sustainable Chemistry: A Critical Review. *ChemSusChem* **2009**, *2*, 1072–1095. [CrossRef]

18. Jawjit, W.; Kroeze, C.; Soontaranun, W.; Hordijk, L. Options to Reduce the Environmental Impact by Eucalyptus-based Kraft Pulp Industry in Thailand: Model Description. *J. Clean. Prod.* **2007**, *15*, 1827–1839. [CrossRef]

19. Bajpai, P. Pulping Fundamentals. In *Biermann's Handbook of Pulp and Paper*; Elsevier: Amsterdam, The Netherlands, 2018; pp. 295–351. ISBN 978-0-12-814240-0.

20. Bajpai, P. Kraft Spent Liquor Recovery. In *Biermann's Handbook of Pulp and Paper*; Elsevier: Amsterdam, The Netherlands, 2018; pp. 425–451. ISBN 978-0-12-814240-0.

21. Haneke, K. *Turpentine (Turpentine Oil, Wood Turpentine, Sulfate Turpentine, Sulfite Turpentine) [8006-64-2] Review of Toxicological Literature*; Integrated Laboratory Systems: Morrisville, NC, USA, 2002.

22. Al-Asheh, S.; Allawzi, M.; Al-Otoom, A.; Allaboun, H.; Al-Zoubi, A. Supercritical Fluid Extraction of Useful Compounds from Sage. *Nat. Sci.* **2012**, *04*, 544–551. [CrossRef]

23. Burdock, G. *Fenaroli's handbook of flavor ingredients*, 6th ed.; CRC Press: Boca Raton, FL, USA, 2010; Volume 14, ISBN 9781420090772.

24. Vickers, C.E.; Behrendorff, J.B.Y.H.; Bongers, M.; Brennan, T.C.R.; Bruschi, M.; Nielsen, L.K. Production of Industrially Relevant Isoprenoid Compounds in Engineered Microbes. In *Microorganisms in Biorefineries*; Springer: Berlin/Heidelberg, Germany, 2015; pp. 303–334. ISBN 978-3-662-45209-7.

25. Meadows, A.L.; Hawkins, K.M.; Tsegaye, Y.; Antipov, E.; Kim, Y.; Raetz, L.; Dahl, R.H.; Tai, A.; Mahatdejkul-Meadows, T.; Xu, L.; et al. Rewriting Yeast Central Carbon Metabolism for Industrial Isoprenoid Production. *Nature* **2016**, *537*, 694–697. [CrossRef]

26. Tashiro, M.; Kiyota, H.; Kawai-Noma, S.; Saito, K.; Ikeuchi, M.; Iijima, Y.; Umeno, D. Bacterial Production of Pinene by a Laboratory-Evolved Pinene-Synthase. *ACS Synth. Biol.* **2016**, *5*, 1011–1020. [CrossRef]

27. Amiri, P.; Shahpiri, A.; Asadollahi, M.A.; Momenbeik, F.; Partow, S. Metabolic Engineering of *Saccharomyces cerevisiae* for Linalool Production. *Biotechnol. Lett.* **2016**, *38*, 503–508. [CrossRef]

28. Bröker, J.N.; Müller, B.; van Deenen, N.; Prüfer, D.; Schulze Gronover, C. Upregulating the Mevalonate Pathway and Repressing Sterol Synthesis in *Saccharomyces cerevisiae* Enhances the Production of Triterpenes. *Appl. Microbiol. Biotechnol.* **2018**, *102*, 6923–6934. [CrossRef]

29. Ignea, C.; Cvetkovic, I.; Loupassaki, S.; Kefalas, P.; Johnson, C.B.; Kampranis, S.C.; Makris, A.M. Improving Yeast Strains using Recyclable Integration Cassettes, for the Production of Plant Terpenoids. *Microb. Cell Fact.* **2011**, *10*, 4. [CrossRef] [PubMed]

30. Ignea, C.; Pontini, M.; Maffei, M.E.; Makris, A.M.; Kampranis, S.C. Engineering Monoterpene Production in Yeast Using a Synthetic Dominant Negative Geranyl Diphosphate Synthase. *ACS Synth. Biol.* **2014**, *3*, 298–306. [CrossRef] [PubMed]

31. Kang, M.-K.; Eom, J.-H.; Kim, Y.; Um, Y.; Woo, H.M. Biosynthesis of Pinene from Glucose using Metabolically-Engineered *Corynebacterium glutamicum*. *Biotechnol. Lett.* **2014**, *36*, 2069–2077. [CrossRef] [PubMed]

32. International Mycological Association *Rhodosporidium toruloide* 2016, Online Database. Available online: http://www.mycobank.org/BioloMICS.aspx?TableKey=14682616000000089&Rec=1019&Fields=All (accessed on 16 December 2019).

33. Zhuang, X.; Kilian, O.; Monroe, E.; Ito, M.; Tran-Gymfi, M.B.; Liu, F.; Davis, R.W.; Mirsiaghi, M.; Sundstrom, E.; Pray, T.; et al. Monoterpene Production by the Carotenogenic Yeast *Rhodosporidium toruloides*. *Microb. Cell Fact.* **2019**, *18*, 54. [CrossRef]

34. Ren, Z.-Y.; Liu, G.-L.; Chi, Z.; Han, Y.-Z.; Hu, Z.; Chi, Z.-M. Overexpression of Both the Lactase Gene and its Transcriptional Activator Gene Greatly Enhances Lactase Production by *Kluyveromyces marxianus*. *Process Biochem.* **2017**, *61*, 38–46. [CrossRef]

35. Nambu-Nishida, Y.; Nishida, K.; Hasunuma, T.; Kondo, A. Development of a Comprehensive Set of Tools for Genome Engineering in a Cold- and Thermo-Tolerant *Kluyveromyces marxianus* Yeast Strain. *Sci. Rep.* **2017**, *7*, 8993. [CrossRef]

36. Bao, S.-H.; Zhang, D.-Y.; Meng, E. Improving Biosynthetic Production of Pinene through Plasmid Recombination Elimination and Pathway Optimization. *Plasmid* **2019**, *105*, 102431. [CrossRef]

37. Niu, F.-X.; Huang, Y.-B.; Ji, L.-N.; Liu, J.-Z. Genomic and Transcriptional Changes in Response to Pinene Tolerance and Overproduction in Evolved *Escherichia coli*. *Synth. Syst. Biotechnol.* **2019**, *4*, 113–119. [CrossRef]

38. Fischer, M.J.C.; Meyer, S.; Claudel, P.; Bergdoll, M.; Karst, F. Metabolic Engineering of Monoterpene Synthesis in Yeast. *Biotechnol. Bioeng.* **2011**, *108*, 1883–1892. [CrossRef]

39. Peng, B.; Nielsen, L.K.; Kampranis, S.C.; Vickers, C.E. Engineered Protein Degradation of Farnesyl Pyrophosphate Synthase is an Effective Regulatory Mechanism to Increase Monoterpene Production in *Saccharomyces cerevisiae*. *Metab. Eng.* **2018**, *47*, 83–93. [CrossRef]

40. Bhattacharya, S.; Esquivel, B.D.; White, T.C. Overexpression or Deletion of Ergosterol Biosynthesis Genes Alters Doubling Time, Response to Stress Agents, and Drug Susceptibility in *Saccharomyces cerevisiae*. *MBio* **2018**, *9*, e01291-18. [CrossRef] [PubMed]

41. Deng, Y.; Sun, M.; Xu, S.; Zhou, J. Enhanced (*S*)-Linalool Production by Fusion Expression of Farnesyl Diphosphate Synthase and Linalool Synthase in *Saccharomyces cerevisiae*. *J. Appl. Microbiol.* **2016**, *121*, 187–195. [CrossRef] [PubMed]

42. Zhao, J.; Li, C.; Zhang, Y.; Shen, Y.; Hou, J.; Bao, X. Dynamic control of ERG20 Expression Combined with Minimized Endogenous Downstream Metabolism Contributes to the Improvement of Geraniol Production in *Saccharomyces cerevisiae*. *Microb. Cell Fact.* **2017**, *16*, 17. [CrossRef] [PubMed]

43. Parulekar, S.; Birol, G.; Cinar, A.; Undey, C. Introduction. In *Batch Fermentation Modeling: Monitoring and Control*; Chemical Industries; CRC Press: Boca Raton, FL, USA, 2003; Volume 93, pp. 1–19. ISBN 978-0-8247-4034-4.

44. Campbel, J.A. Energy from Aerobic vs. Anaerobic Fermentation. *J. Chem. Educ.* **1973**, *50*, 535.

45. Abram, T. *Sugars Defined*; Michigan State University Extension: Lansing, MI, USA, 2014.

46. Hughes, P.; Risner, D.; Goddik, L.M. Whey to Vodka. In *Whey-Biological Properties and Alternatives*; IntechOpen: London, UK, 2018; Available online: https://www.intechopen.com/books/whey-biological-properties-and-alternative-uses/whey-to-vodka (accessed on 16 December 2019).

47. Tetra Pak Whey Processing. In *Dairy Processing Handbook*; Tetrapak, 2019; Available online: https://dairyprocessinghandbook.tetrapak.com/chapter/whey-processing (accessed on 16 December 2019).

48. Lee, H.; Cuthbertson, D.J.; Otter, D.E.; Barile, D. Rapid Screening of Bovine Milk Oligosaccharides in a Whey Permeate Product and Domestic Animal Milks by Accurate Mass Database and Tandem Mass Spectral Library. *J. Agric. Food Chem.* **2016**, *64*, 6364–6374. [CrossRef] [PubMed]

49. Dallas, D.C.; Weinborn, V.; de Moura Bell, J.M.L.N.; Wang, M.; Parker, E.A.; Guerrero, A.; Hettinga, K.A.; Lebrilla, C.B.; German, J.B.; Barile, D. Comprehensive Peptidomic and Glycomic Evaluation Reveals that Sweet Whey Permeate from Colostrum is a Source of Milk Protein-Derived Peptides and Oligosaccharides. *Food Res. Int.* **2014**, *63*, 203–209. [CrossRef]

50. Siso, M.I.G. The biotechnological utilization of cheese whey: A review. *Bioresour. Technol.* **1996**, *57*, 1–11. [CrossRef]

51. Pasotti, L.; Zucca, S.; Casanova, M.; Micoli, G.; Cusella De Angelis, M.G.; Magni, P. Fermentation of Lactose to Ethanol in Cheese Whey Permeate and Concentrated Permeate by Engineered *Escherichia coli*. *BMC Biotechnol.* **2017**, *17*, 48. [CrossRef]

52. Risner, D.; Tomasino, E.; Hughes, P.; Meunier-Goddik, L. Volatile Aroma Composition of Distillates Produced from Fermented Sweet and Acid Whey. *J. Dairy Sci.* **2019**, *102*, 202–210. [CrossRef]

53. Akbas, M.Y.; Sar, T.; Ozcelik, B. Improved Ethanol Production from Cheese Whey, Whey Powder, and Sugar Beet Molasses by " *Vitreoscilla* hemoglobin expressing" *Escherichia coli*. *Biosci. Biotechnol. Biochem.* **2014**, *78*, 687–694. [CrossRef]

54. Sar, T.; Stark, B.C.; Yesilcimen Akbas, M. Effective Ethanol Production from Whey Powder through Immobilized *E. coli* Expressing *Vitreoscilla* Hemoglobin. *Bioengineered* **2017**, *8*, 171–181. [CrossRef] [PubMed]

55. Guimarães, P.M.R.; François, J.; Parrou, J.L.; Teixeira, J.A.; Domingues, L. Adaptive Evolution of a Lactose-consuming *Saccharomyces cerevisiae* recombinant. *Appl. Environ. Microbiol.* **2008**, *74*, 1748–1756. [CrossRef] [PubMed]

56. Sreekrishna, K.; Dickson, R.C. Construction of Strains of *Saccharomyces cerevisiae* that Grow on Lactose. *Proc. Natl. Acad. Sci. USA* **1985**, *82*, 7909–7913. [CrossRef] [PubMed]

57. Lawton, M.R.; Alcaine, S.D. Leveraging Endogenous Barley Enzymes to Turn Lactose-Containing Dairy By-products into Fermentable Adjuncts for *Saccharomyces cerevisiae*-based Ethanol Fermentations. *J. Dairy Sci.* **2019**, *102*, 2044–2050. [CrossRef]

58. Liu, J.-J.; Zhang, G.-C.; Oh, E.J.; Pathanibul, P.; Turner, T.L.; Jin, Y.-S. Lactose Fermentation by Engineered *Saccharomyces cerevisiae* Capable of Fermenting Cellobiose. *J. Biotechnol.* **2016**, *234*, 99–104. [CrossRef]

59. Barrett, E.; Stanton, C.; Zelder, O.; Fitzgerald, G.; Ross, R.P. Heterologous Expression of Lactose- and Galactose-utilizing Pathways from Lactic Acid Bacteria in *Corynebacterium glutamicum* for Production of Lysine in Whey. *Appl. Environ. Microbiol.* **2004**, *70*, 2861–2866. [CrossRef] [PubMed]

60. Akhtar, P.; Gray, J.I.; Asghar, A. Synthesis of Lipids by Certain Yeast Strains Grown on Whey Permeate. *J. Food Lipids* **1998**, *5*, 283–297. [CrossRef]

61. Banat, I.M.; Nigam, P.; Marchant, R. Isolation of Thermotolerant, Fermentative Yeasts Growing at 52 °C and Producing Ethanol at 45 °C and 50 °C. *World J. Microbiol. Biotechnol.* **1992**, *8*, 259–263. [CrossRef]
62. Groeneveld, P.; Stouthamer, A.H.; Westerhoff, H.V. Super Life—How and Why 'Cell Selection' Leads to the Fastest-Growing Eukaryote. *FEBS J.* **2009**, *276*, 254–270. [CrossRef]
63. Cernak, P.; Estrela, R.; Poddar, S.; Skerker, J.M.; Cheng, Y.-F.; Carlson, A.K.; Chen, B.; Glynn, V.M.; Furlan, M.; Ryan, O.W.; et al. Engineering *Kluyveromyces marxianus* as a Robust Synthetic Biology Platform Host. *MBio* **2018**, *9*, e01410-18. [CrossRef]
64. Ibdah, M.; Muchlinski, A.; Yahyaa, M.; Nawade, B.; Tholl, D. *Carrot Volatile Terpene Metabolism: Terpene Diversity and Biosynthetic Genes*; Springer: Cham, Switzerland, 2019; pp. 279–293.
65. Duarte, S.H.; dos Santos, P.; Michelon, M.; de Pinho Oliveira, S.M.; Martínez, J.; Maugeri, F. Recovery of yeast lipids using Different Cell Disruption Techniques and Supercritical CO_2 Extraction. *Biochem. Eng. J.* **2017**, *125*, 230–237. [CrossRef]
66. Reverchon, E. Supercritical Fluid Extraction and Fractionation of Essential Oils and Related Products. *J. Supercrit. Fluids* **1997**, *10*, 1–37. [CrossRef]

Article

Comparative Study of Angiotensin I-Converting Enzyme (ACE) Inhibition of Soy Foods as Affected by Processing Methods and Protein Isolation

Cíntia L. Handa [1], Yan Zhang [2], Shweta Kumari [3], Jing Xu [4], Elza I. Ida [5] and Sam K. C. Chang [6,*]

[1]   Minas Gerais State University, R. Ver. Geraldo Moisés da Silva, 308-434—Universitário, Ituiutaba 38302-182, MG, Brazil; cintia.handa@uemg.br
[2]   Coastal Research and Extension, Mississippi State University, Starkville, MS 39762, USA; yzhang@fsnhp.msstate.edu
[3]   Department of Nutrition, University of Texas at San Antonio, San Antonia, TX 78023, USA; shweta.kumari@utsa.edu
[4]   College of Science, Northeast Agricultural University, Harbin 150030, China; xujing@neau.edu.cn
[5]   Departamento de Ciência e Tecnologia de Alimentos, Universidade Estadual de Londrina, Londrina 86057-970, PR, Brazil; elida@uel.br
[6]   Coastal Research and Extension Center, Department of Food Science Nutrition and Health Promotion, Mississippi State University, Starkville, MS 39762, USA
*   Correspondence: sc1690@msstate.edu; Tel.: +1-228-762-7783; Fax: +1-228-388-7493

Received: 14 July 2020; Accepted: 7 August 2020; Published: 12 August 2020

Abstract: Angiotensin converting enzyme (ACE) converts angiotensin I into the vasoconstrictor angiotensin II and eventually elevates blood pressure. High blood pressure is a major risk factor for heart disease and stroke. Studies show peptides present anti-hypertensive activity by ACE inhibition. During food processing and digestion, food proteins may be hydrolyzed and release peptides. Our objective was to determine and compare the ACE inhibitory potential of fermented and non-fermented soy foods and isolated 7S and 11S protein fractions. Soy foods (e.g., soybean, natto, tempeh, yogurt, soymilk, tofu, soy-sprouts) and isolated proteins were in vitro digested prior to the determination of ACE inhibitory activity. Peptide molecular weight distribution in digested samples was analyzed and correlated with ACE inhibitory capacity. Raw and cooked soymilk showed the highest ACE inhibitory potential. Bacteria-fermented soy foods had higher ACE inhibitory activity than fungus-fermented soy food, and 3 day germinated sprouts had higher ACE inhibition than those germinated for 5 and 7 days. The 11S hydrolysates showed higher ACE inhibitory capacity than 7S. Peptides of 1–4.5 kDa showed a higher contribution to reducing IC_{50}. This study provides evidence that soy foods and isolated 7S and 11S proteins may be used as functional foods or ingredients to prevent or control hypertension.

Keywords: peptides; hydrolysates; hypertension

---

1. Introduction

High blood pressure is a major risk factor for heart attacks and strokes [1]. Angiotensin converting enzyme (ACE) (peptidyl-dipeptidase, E.C. 3.4.15.1) plays a vital role in the regulation of blood pressure in the renin–angiotensin system. Inhibition of ACE is the dominant therapeutic approach to treat high blood pressure, since angiotensin II generation is inhibited and bradykinin is preserved. In this context, research is being developed in order to study the potential food-derived ACE inhibitors [2].

Soybean (*Glycine max* (L.) Merrill) is rich in proteins, and 11S and 7S proteins are two major proteins representing more than 70% of the total proteins [3] with the rest being lipoxygenases, trypsin inhibitors, lectins, lunasin, and other minor proteins. Soy isolates are composed mostly of 11S and 7S proteins and are used as protein ingredients in many processed food products. Peptides from soybean have been identified and characterized for their potential ACE inhibition [4,5]. Inhibitory peptides of ACE have been reported to exist in soy protein hydrolysates [6,7], fermented soy foods, such as soybean paste [8], tempeh and natto [6], and in non-fermented soy foods such as soybean-based infant formulas and soymilk [4,9,10].

Peptides and phenolic compounds released during food processing or in vitro digestion usually show a multifunctional nature including antioxidant, immunomodulatory, antimicrobial, antithrombotic, hypocholesterolemic, and anti-hypertensive potential [11]. However, foods must be digested before absorption can occur. Therefore, it is meaningful to conduct simulated in vitro digestion prior to the assay for ACE inhibition.

Many food processing technologies, employing various physical, chemical, enzymatic, biological, and engineering principles or a combination of these are used in making various fermented (natto, tempeh, and soy yogurt) and non-fermented soy foods or protein ingredients (soy isolates, protein hydrolysates, soymilk, tofu, and soy sprouts). The processing technologies have a major influence on the peptides and phenolic contents and compositions in the products.

In addition to the variability of processing methods, there are hundreds of soybean varieties with different protein and phenolic compositions. Therefore, due to the differences in raw materials, food processing methods and storage conditions, the ability of ACE inhibition of soy products reported in the literature is very difficult to compare, since the reported studies used different materials and methods. To attempt to understand how food processing and storage methods affect ACE inhibition of the products, the same soybean variety must be used. Thus far, a systematic study to compare the ACE inhibition of various commonly consumed soy products using one soybean variety is lacking.

Therefore, the objectives of this study were: (i) to investigate the ACE inhibitory activity of in vitro digested soy foods, such as soymilk, tofu, sprout, natto, tempeh and soy yogurt, made from the same soybean variety; (ii) to evaluate the effect of hydrolysis of 7S and 11S protein fractions on ACE inhibitory capacity; and (iii) to determine the correlation between ACE inhibitory capacity and peptide molecular size.

2. Materials and Methods

2.1. Soybean Material and Chemicals

The soybean cultivar *Prosoy*, harvested in 2012, was obtained from Sinner Brothers and Bresnahan Co. grown in Casselton, ND, USA. Raw soybean (RSoy) was stored in a cool and dry air-conditioned room (5 °C) prior to use.

Pepsin from porcine gastric mucosa, pancreatin from porcine pancreas, trypsin from porcine pancreas, α-chymotrypsin from bovine pancreas, captopril, hippuric acid (HA), ACE, N-α-hippuryl-L-histidyl-L-leucine (HHL), HPLC-grade trifluoroacetic acid (TFA) were purchased from Sigma–Aldrich (St. Louis, MO, USA).

2.2. Preparation of Soy Foods

2.2.1. Soymilk

Raw traditional soymilk (RTSoyM), cooked traditional soymilk (CTSoyM), raw cooked soymilk slurry (RSoyMS), and cooked soymilk slurry (CSoyMSF) were prepared. RTSoyM was prepared according to Zhang et al. [12]. For CSoyMSF, the soy slurry was subjected to the same cooking method as that used for soymilk (obtained by filtration first), followed by filtration through muslin cloth to separate the okara from the soymilk. Both types of soymilk (filtered the slurry then cooked, and cooked the slurry then filtered) were cooled in an ice bath, freeze-dried, and stored at −20 °C before analysis.

2.2.2. Tofu

Pressed tofu (PT) and filled tofu (FT) were prepared according to Meng et al. [13]. Both PT and FT were freeze-dried and stored at −20 °C before analysis.

2.2.3. Natto

Natto was prepared as reported by our previous study [14]. *Bacillus natto* culture was prepared from a commercial natto purchased from an Asian grocery store in Starkville, MS, USA. Natto (N) samples were stored at 4 °C for 2 (Nd2), 4 (Nd4), and 6 days (Nd6), then freeze-dried and stored at −20 °C before analysis.

2.2.4. Tempeh

Tempeh was prepared using starter culture, *Rhizopus oryzae* obtained from the Cultures for Health (Morrisville, NC, USA). Manufacturer's recommended protocol was followed with minor modifications. Soybeans were soaked in water at 30 °C until pH reached 6.5, and were then washed, steamed for 15 min and dried on a mesh screen with an electric fan to evaporate the water of the beans. Soybeans were inoculated with the culture and packaged in zip lock bags with tiny holes that were manually poked with toothpicks. The sample was incubated at 31 °C for 48 h. Tempeh (T) samples were kept at 4 °C for 2 (Td2), 4 (Td4), and 6 days (Td6) before freeze-drying and then stored at −20 °C before analysis.

2.2.5. Soy Yogurt

Soy yogurt was made as follows: To make soymilk, the bean-to-water ratio used was 1:7 (*w/w*). Soymilk was sterilized at 121 °C for 5 min in glass bottles. When soymilk reached 40 °C, aseptic inoculation was conducted according to the manufacturer's recommended rate with a yogurt culture (YC−087; Chr. Hansen Laboratory, Inc., Milwaukee, WI, USA) that contained strains of *Streptococcus thermophilus* and *L. delbrueckii* ssp. bulgaricus. The inoculated soymilk was poured into 500 mL sterile transparent plastic cups with lids (300 mL per cup) and incubated at 40 °C until the pH decreased to between 4.2 and 4.5. The soy yogurt (Y) samples were stored at 4 °C for 0 (Yd0), 2(Yd2), 6 (Yd6), and 8 days (Yd8) before freeze-drying and then stored at −20 °C before analysis.

2.2.6. Soybean Sprout

Soybean sprouts were prepared as reported by Kumari and Chang [15] using Freshlife Automatic Sprouter (Tribest Corporation, Cerritos, CA, USA). Raw sprouts (RS) germinated for 1, 2, 3, 5, and 7 days were designated as RSd1, RSd2, RSd3, RSd5, and RSd7, respectively. Meanwhile, a portion of each type of sprout was cooked using common household cooking practice according to Kumari and Chang [15]. The cooked sprouts (CS) were named as CSd1, CSd2, CSd3, CSd5, and CSd7, respectively. All sprouts were freeze-dried for further analysis.

2.3. Simulated In Vitro Digestion of Soy Foods

The freeze-dried soy foods were subjected to simulated in vitro gastrointestinal digestion using pepsin and pancreatin in sequence (done in duplicate). Freeze-dried soy foods (3 g) were put in a 50 mL conical flask containing 30 mL of distilled water and mixed for 30 s. For each sample, 0.3 mL of pepsin (10,000 units/mL) was added, mixed, and the pH was adjusted to 2.0 using 5 N HCl, before the flasks were incubated at 37 °C in a shaking water bath (220 rpm) for 2 h. The pH of the pepsin digests was adjusted to 8.0 using 5 N NaOH and 3 mL of pancreatin (40 mg/mL) was added to each flask. The pH was adjusted again to 8.0 using 5 N NaOH. Incubation was continued for 2 h. The digested sample was held in the boiling water for 15 min to inactivate the digestive enzymes, and then it was cooled and centrifuged (3000× *g*, 15 min). The digested sample was filtered, freeze-dried and stored at −20 °C until use. The freeze-dried digested samples were suspended in water and adjusted to 0.25 mg/mL (on dried basis) for ACE inhibitory activity.

2.4. Characterization of Digested Soy Foods

From the results of ACE inhibitory activity of all soy foods subjected to simulated in vitro digestion, the digested soy foods for each type of processing methods that gave a higher percentage of ACE inhibition were selected to represent that category of processing methods for further characterizations of ACE inhibitory capacity (IC_{50} = concentration of extracts required to inhibit 50% of the enzyme activity), soluble proteins, and peptide molecular weight (MW) distribution.

2.5. Preparation and Hydrolysis of 7S and 11S Fractions

To study the ACE inhibition effect of the major storage proteins, 7S and 11S fractions were isolated from low-temperature defatted soybean meal according to the method of Liu et al. [16] and were freeze-dried and stored at −20 °C. Before hydrolysis, to remove the effect of residual phenolic compounds in the soy proteins, the freeze-dried 7S and 11S fractions were suspended into 80% acetone, shaken overnight at 25 °C and centrifuged at 15,000× g for 30 min. The precipitates were dried in an oven at 40 °C for 48 h.

To study the relationships of the peptides produced after each digestive protease, and their resultant ACE inhibition potency, duplicated experiments were carried out in five steps with steps B, C, and D to study the stepwise effect of pepsin, trypsin, and α-chymotrypsin sequentially, and step E to study the combined effect of the two pancreatic proteases (trypsin and α-chymotrypsin) following pepsin digestion: (A) 7S and 11S fractions were suspended in deionized water (10 g into 100 mL); (B) the pH value was adjusted to 2 with 5 N HCl, 1 mL of pepsin (10,000 units/mL) was added, and the suspension was incubated at 37 °C in a shaking water bath (220 rpm) for 2 h; (C) the pH was adjusted to 8 using 5 N NaOH, 0.7 mL of trypsin (520,000–800,000 units/mL) was added, and incubation was continued for 2 h; (D) the pH was adjusted to 8 using 5 N NaOH, 0.4 mL of α-chymotrypsin (1600 units/mL) was added, and incubation was continued for additional 2 h; and (E) 7S and 11S fractions were suspended in deionized water (3 g into 30 mL in a flask), the pH value was adjusted to 2.0 with 5 N HCl, 0.3 mL of pepsin (10,000 units/mL) was added, and the flasks incubated at 37 °C in a shaking water bath (220 rpm) for 2 h; then the pH was adjusted to 8 using 5 N NaOH, 0.3 mL of trypsin (520,000–800,000 units/mL) and 0.3 mL of α-chymotrypsin (1600 units/mL) were added, and incubation was continued for 2 h. Aliquots of 30 mL were collected from the reaction mixtures after the completion of (B), (C), (D), and (E) steps, respectively. The collected mixtures were heated at 100 °C for 10 min to inactivate enzymes, and were labeled as follows: A—no hydrolyzed; B—hydrolyzed by pepsin for 2 h; C—hydrolyzed by pepsin for 2 h + trypsin for 2 h; D—hydrolyzed by pepsin for 2 h + trypsin for 2 h + α-chymotrypsin for 2 h; and E—hydrolyzed by pepsin for 2 h + (trypsin and α-chymotrypsin) for 2 h. The hydrolysates were cooled, freeze-dried and stored at −20 °C until use. These hydrolysates were suspended in water (10 mg/mL, on dried basis) for measuring ACE inhibitory capacity (IC_{50}) and peptide MW distribution.

2.6. Determination of Angiotensin Converting Enzyme (ACE) Inhibitory Activity

In vitro ACE inhibitory activity was measured according to Cushman and Cheung [17] with modifications. N-α-hippuryl-L-histidyl-L-leucine (HHL) and ACE were dissolved in 100 mM borate buffer (pH = 8.3, 300 mM NaCl), at concentrations of 2.5 mM and 10 mU/mL, respectively. The reaction mixture containing 75 μL HHL, 35 μL sample, and 75 μL ACE was incubated at 37 °C in a shaking water bath (200 rpm) for 30 min. The reaction was stopped by heating at 85 °C for 15 min and 185 μL of water was added before injecting into the ultra-high-performance liquid chromatography (UHPLC) sample loop. The UHPLC analyses were performed to quantify the hippuric acid (HA) produced by the enzymatic hydrolysis of the HHL using a Thermo Scientific Dionex UltiMate 3000 RSLC UHPLC focused System (Thermo Scientific Dionex, Fürstenfeldbruck, Germany). Data acquisition was performed with Chromeleon 7 software. Separations were accomplished on a Waters UPLC column (CORTECS® UPLC®, C18 1.6 μm, 2.1 × 50 mm) and column temperature was maintained at 37 °C. The analytical procedure for quantification of HA by UHPLC was previously validated. The injection volume was 100 μL and the detection wavelength was set at 228 nm. The mobile phase consisted of a gradient of 0.1% TFA in water (A) and 0.1% TFA in acetonitrile (B). The flow rate was set at 0.176 mL/min, and the gradient profile was as follows: t0–0.96 min: linear gradient from 5% to 30% of B, and 95% to 70% of A; t0.96–1.60 min: linear gradient from 30% to 60% of B, and 70% to 40% of A; t1.60–1.90 min: curve 7 gradient from 60% to 70% of B, and 40% to 30% of A; t1.90–2.2 0 min: isocratic elution with 70% of B, and 30% of A; t2.20–2.60 min: linear gradient from 70% to 5% of B, and 30% to 95% of A; total time of 4 min. The HA control contained water instead of sample solution, and the blank sample contained buffer instead of ACE solution. The ACE inhibitory activity was calculated according to the following equation:

$$\text{ACE inhibitory activity (\%)} = ((A - B)/(A)) \times 100, \tag{1}$$

where A is HA from control containing water instead of sample solution and B is HA from sample reaction with the subtraction of sample blank. The concentration to inhibit ACE by 50 % (IC_{50}) (mg/mL on dried basis) was calculated using GraphPad Prism software.

2.7. Determination of Peptide Molecular Weight Distribution

Molecular Weight Distribution of peptides from digested samples was analyzed by size exclusion chromatography [18] with modifications. Each sample was dissolved in water, at a concentration of 10 mg/mL (on dried basis) for 7 S and 11S hydrolysates, and 8.6 mg/mL (on dried basis) for digested soy foods. A 40 μL aliquot was injected onto the Superdex™ peptide 10/300 GL column (GE Healthcare, Piscataway, NJ, USA) coupled with an Agilent Technologies Chromatography 1200 series system (Agilent Technologies, Santa Clara, CA, USA), and eluted with 30% acetonitrile at a flow rate of 0.380 mL/min. The eluate was monitored at 214 nm. A calibration curve of MW was obtained with the following standards: cytochrome C (12,384 Da), aprotinin (6500 Da), vitamin B12 (1855 Da), and L-reduced glutathione (307 Da).

2.8. Determination of Soluble Proteins

Soluble proteins were measured by bicinchoninic acid (BCA) method [19] and expressed as micrograms/milligram of sample (μg/mg) on dried basis with bovine serum albumin (BSA) as a standard.

2.9. Statistical Analysis

The soy food preparation, in vitro gastrointestinal digestion and analyses were carried out in duplicate, duplicate and triplicate, respectively. One-way analysis of variance (ANOVA) tests followed by Tukey's multiple comparisons test ($\alpha = 0.05$) was carried out using the software program Statistic 10 (StatSoft, Tulsa, OK, USA). Differences among sample groups were visualized by principal component analysis (PCA).

3. Results and Discussion

3.1. Angiotensin Converting Enzyme (ACE) Inhibitory Activity of Digested Fermented and Non-Fermented Soy Foods

The processing methods employed to manufacture soy foods, such as cooking, grinding, soaking, coagulation, germination, dilution, and fermentation, influenced the ACE inhibitory activity of the digested soy foods (Figure 1). Soymilk, tofu, and raw sprout germinated for 1, 2, and 3 days, and most fermented soy foods (all natto and yogurt products and Tempeh 2 day) had increased in % ACE inhibition as compared to raw soybean (RSoy). However, raw sprout germinated for 7 days (RSd7), cooked sprouts and tempeh stored for 6 days had lower % ACE inhibition than RSoy. The influence of the thermal treatment on ACE inhibitory activity was dependent upon the type of soy food as observed in the raw and cooked traditional soymilk (RTSoyM and CTSoyM), whose ACE inhibition percentage did not distinguish from each other. While the CSoyMSF (soymilk from filtered cooked slurry) showed a 14% reduction in the ACE inhibitory activity as compared to RSoyMS (raw soymilk slurry). The greatest impact of heat treatment on ACE inhibitory activity of sprouts was observed in the cooked sprout germinated for 3 days (CSd3), which showed a 69% reduction in comparison with its correspondent raw sprout (RSd3).

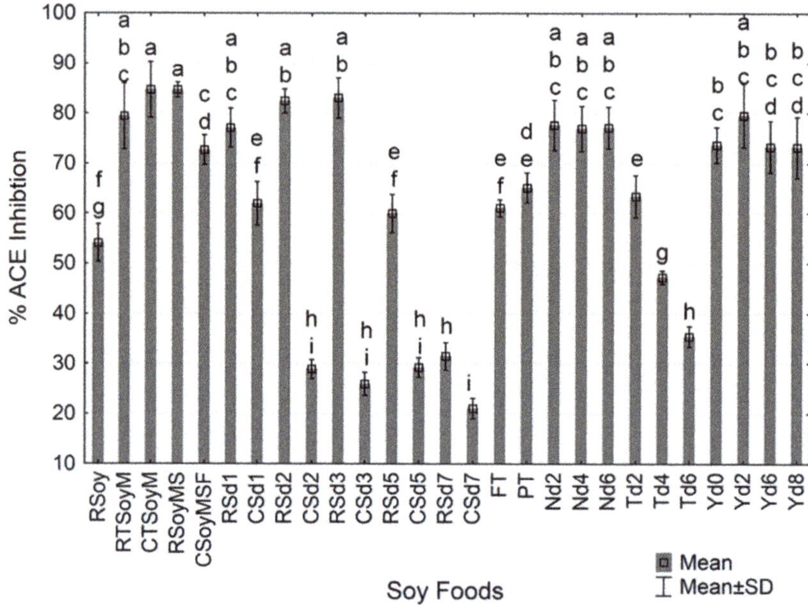

Figure 1. Percentage inhibition of angiotensin converting enzyme (ACE) of digested fermented and non-fermented soy foods. Results are expressed as the mean ± standard deviation. Sample concentration = 0.25 mg/mL in water (on dried basis). RSoy: raw soybean; RTSoyM: raw traditional soymilk; CTSoyM: cooked traditional soymilk; RSoyMS: raw soymilk slurry; CSoyMSF: cooked soymilk with okara and filtered; PT: pressed tofu; FT: filled tofu; RSd (1, 2, 3, 5, and 7): raw sprout germinated for 1, 2, 3, 5, and 7 days; CSd (1, 2, 3, 5, and 7): cooked sprout germinated for 1, 2, 3, 5, and 7 days; Nd (2, 4, and 6): natto stored for 2, 4, and 6 days; Td (2, 4, and 6): tempeh stored for 2, 4, and 6 days; and Yd (2, 4, 6, and 8): yogurt stored for 2, 4, and 6 days. Different lowercase letters over the bar are significantly ($p < 0.05$) different among samples.

The thermal treatments applied in the preparation of the different soy foods, possibly, altered the protein structure in different ways, affecting the enzyme activity or hydrolysis sites on the peptide chains during in vitro digestion. Consequently, different peptides may be formed [20].

Germination time also significantly affected the ACE inhibition percentage of the sprouts with a 1.5 fold increase after 24 h of germination. The ACE inhibition remained constant through the third day and then decreased until the seventh day which showed about 42% smaller inhibition than RSoy. The increase in the % ACE inhibition followed by reduction could be attributed to the excessive hydrolysis of the proteins or to the breakdown of the complex phenolic compounds caused by long germination time followed by in vitro digestion. According to Zakharov et al. [21], at the beginning of germination, the proteins started to become hydrolyzed by endopeptidases to oligopeptides, and then were hydrolyzed by exopeptidases to free amino acids which could be used to synthesize new proteins and tissues. Yang and Li [22] found by sodium dodecyl sulfate polyacrylamide gel electrophoresis (SDS-PAGE) that α' and α subunits of β-conglycinin (7S) and the acidic chains of glycinin (11S) were gradually degraded with germination time right after the soybean imbibition. During gastrointestinal digestion, the additional presence of exopeptidases and endopeptidases would further degrade these compounds [4].

The angiotensin converting enzyme inhibitory activity was dependent on the type of microorganisms used in the making of fermented soy foods. Bacteria and fungi were used in the solid-state fermentation to obtain natto and tempeh, respectively. Whereas, soy yogurt was fermented with bacteria by submersion fermentation. As observed in Figure 1, natto stored for 2 days (Nd2) showed an increase of 40% in the ACE inhibitory activity as compared to RSoy and remained constant until the sixth day of storage. Tempeh stored for 2 days (Td2) increased just 17% in the ACE inhibitory activity as compared to RSoy, and the activity decreased with the storage time. However, the ACE inhibitory activity of soy yogurt was 36% higher than RSoy and about 11% smaller than CTSoyM (cooked traditional soymilk) and remained unchanged during the storage.

The differences in the ACE inhibitory activity between natto and tempeh could be due to the presence of different enzymes, since fungi usually produce a wider range of extracellular enzymes than bacteria [23]. The proteases from *Bacillus* and *Rhizopus* strains may hydrolyze the main soy proteins into large peptides [8], and the subsequent hydrolysis of these peptides during in vitro digestion by digestive enzymes could lead to formation of different, smaller peptides. Therefore, the high % ACE inhibition and storage stability of natto and soy yogurt compared to tempeh indicated that bacteria could be more promising than fungi for production of fermented soy foods with better ACE inhibitory activity.

3.2. Characterization of Digested Soy Foods

Table 1 and Figure 2 show the ACE inhibitory capacity (IC_{50}), soluble proteins, and peptide MW distribution of the representative digested soy foods that gave higher percentages of ACE inhibition in each food processing category. As shown in Table 1, soymilk exhibited the highest ACE inhibition with RTSoyM (raw traditional soymilk) and CTSoyM (cooked traditional soymilk) representing approximately 47% and 37% IC_{50} decreases, respectively, in comparison with RSoy (raw soybean).

Table 1. Characteristics of the soy products after simulated in vitro digestion.

Digested Soy Foods	IC_{50} (mg/mL)	Soluble Protein (μg/mg)
RSoy	0.19 ± 0.01 [d]	368.64 ± 4.25 [d]
RTSoyM	0.10 ± 0.01 [a]	546.36 ± 2.04 [a,b]
CTSoyM	0.12 ± 0.01 [a,b]	482.38 ± 18.82 [a,b,c,d]
PT	0.20 ± 0.00 [d]	432.00 ± 15.31 [b,c,d]
RSd3	0.15 ± 0.00 [b,c]	562.46 ± 10.91 [a]
CSd3	0.20 ± 0.02 [d]	381.32 ± 15.81 [c,d]
Nd2	0.16 ± 0.01 [c,d]	569.60 ± 8.98 [a]
Td2	0.26 ± 0.00 [e]	487.44 ± 2.57 [a,b,c]
Yd2	0.17 ± 0.01 [c,d]	418.19 ± 8.07 [c,d]

Results are expressed as the mean ± standard deviation (on dried basis). Values followed by different superscript lowercase letters in the same column differ significantly ($p < 0.05$). IC_{50}: concentration to inhibit angiotensin converting enzyme (ACE) by 50 %; RSoy: raw soybean; RTSoyM: raw traditional soymilk; CTSoyM: cooked traditional soymilk; PT: pressed tofu; RSd3: raw sprout germinated for 3 days; CSd3: cooked sprout germinated for 3 days; Nd2: natto stored for 2 days; Td2: tempeh stored for 2 days; and Yd2: yogurt stored for 2 days.

However, when the soymilk was fermented into soy yogurt or made into pressed tofu (PT), the IC_{50} values were not significantly different from that of raw soybean ($p > 0.05$). The effect of solid-state fermentation on ACE inhibitory capacity was dependent on the specific type of microorganisms used in the making of soy foods. IC_{50} value of the natto 2 day (Nd2, made from *Bacillus natto*) was not significantly different from that of RSoy (raw soybean), whereas Td2 (made from *Rhizopus* species) showed a 1.36 fold increase in the IC_{50} value as compared with RSoy. The sprouting of soybean for 3 days (RSd3) reduced about 21% of the IC_{50} value in comparison with RSoy, but the cooking of sprout (CSd3) increased the IC_{50} value which was equivalent to that of RSoy thus eliminating the advantage of sprouting.

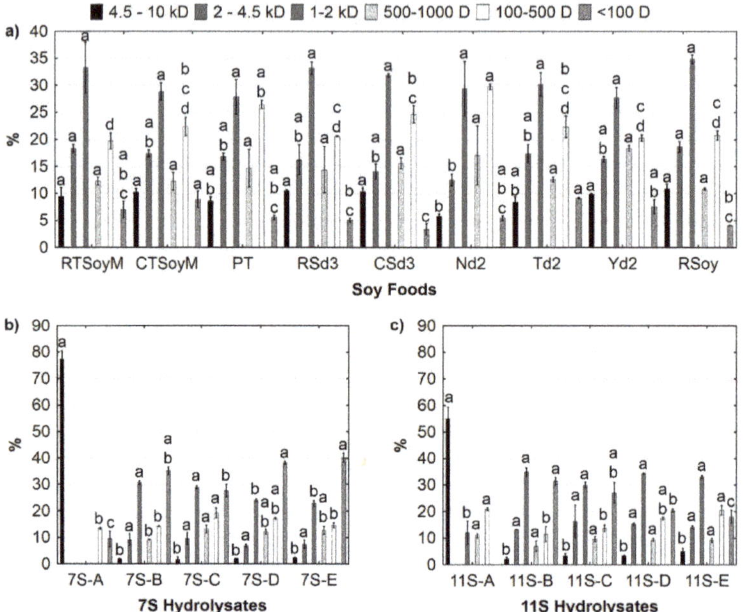

Figure 2. Molecular weight distribution of peptides from the hydrolysates of the simulated in vitro digested soy foods (**a**), 7S protein (**b**), and 11S protein (**c**). RSoy: raw soybean; RTSoyM: raw traditional soymilk; CTSoyM: cooked traditional soymilk; PT: pressed tofu; RSd3: raw sprout germinated for 3 days; CSd3: cooked sprout germinated for 3 days; Nd2: natto stored for 2 days; Td2: tempeh stored for 2 days; and Yd2: yogurt stored for 2 days. 7S and 11S: A—no hydrolyzed; B—hydrolyzed by pepsin for 2 h; C—hydrolyzed by pepsin for 2 h + trypsin for 2h; D—hydrolyzed by pepsin for 2 h + trypsin for 2 h + α-chymotrypsin for 2 h; and E—hydrolyzed by pepsin for 2 h + trypsin and α-chymotrypsin for 2 h. Different lowercase letters over the bar denote significant differences among samples ($p < 0.05$) on the same range of peptide molecular weight.

Comparing the IC_{50} values from different digested fermented and non-fermented soy foods from the seeds of the same soybean variety had not been reported in the literature. Our research agreed with the concept that ACE inhibition ability depended on the experimental protocol, protein extraction procedure, and differences in the peptide mixture compositions after digestion process [24].

The IC_{50} value of the synthetic inhibitor, captopril, was analyzed in our laboratory to verify ACE inhibition efficiency of the digested soy foods. Results showed captopril was a very strong inhibitor as compared with digested soy foods. The IC_{50} value of captopril was 1.05 nM, which was comparable to that reported by Lahogue et al. [25]. Our research is consistent with the report of Hayes and Tiwari [26] that bioactive peptides derived from natural sources needed higher concentrations than synthetic drugs to be effective.

The samples Nd2, Td2, RTSoyM, CTSoyM, and RSd3 showed relatively higher soluble proteins, while RSoy gave the lowest soluble proteins (Table 1) which reflected protein digestibility. Raw soy contains trypsin inhibitors and the native protein structure was compact and resistant to digestion by proteases. The soluble proteins content was influenced by heat treatment the same way as observed in the ACE inhibitory capacity. After the size exclusion chromatography, the separated peptides were grouped into six MW ranges: 4.5–10, 2–4.5, 1–2, 0.5–1, 0.1–0.5, and <0.1 kDa. Most of peptides (>70%) present in soy foods exhibited MW below 2 kDa, although the 1–2 kDa group was pre-dominant in the most of samples. As shown in Figure 2a, the processing had little effect on peptide MW distribution. Except Nd2, which showed about 47% and 43% lower than Rsoy in the 4.5–10 and 2–4.5 kDa ranges, respectively, all other soy foods exhibited similar MW distribution.

The relative content of peptides ranging from 1–2 and 0.5–1 kDa were similar among digested soy foods ($p > 0.05$). The highest percentages between 0.1 and 0.5 kDa were observed in the Nd2 and PT, which were 1.43 and 1.27-fold higher than RSoy, respectively, while no differences existed in the relative content of these peptides between other soy foods and RSoy. CTSoyM and Td2 showed higher percentages of compounds <0.1 kDa than RSoy. The compounds less than 0.5 kDa could be amino acids released by the simulated in vitro digestion. In addition to the protein structural factors (such as degrees of denaturation, unfolding and aggregations) that were induced by processing conditions, the differences in the peptide MW distribution could be due to the lack of specificity of the digestive enzymes and of the proteases which naturally present in different types of foods, particularly fermented foods [4].

3.3. Angiotensin Converting Enzyme (ACE) Inhibitory Capacity and Molecular Weight Distribution of 7S and 11S Hydrolysates

As shown in Table 2, each enzyme sequentially involved changed the ACE inhibitory capacity (A—no hydrolyzed; B—hydrolyzed by pepsin for 2 h; C—hydrolyzed by pepsin for 2 h + trypsin for 2 h; D—hydrolyzed by pepsin for 2 h + trypsin for 2 h + α-chymotrypsin for 2 h; and E—hydrolyzed by pepsin for 2 h + trypsin and α-chymotrypsin for 2 h), and the two major storage proteins also exhibited significant ($p < 0.05$) differences. The IC$_{50}$ value of non-hydrolyzed 7S-A fraction was 2.16 fold higher than that of non-hydrolyzed 11S-A. Pepsin hydrolysis (B) for 2 h of 7S and 11S fractions reduced the IC$_{50}$ value by about 32% and 24%, respectively. The IC$_{50}$ values of 7S and 11S hydrolyzed by pepsin for 2 h + trypsin for 2 h (C) were approximately 81% and 84% smaller than those of 7S-A and 11S-A, respectively, and remained constant with addition of α-chymotrypsin for 2 h (D). No significant differences were observed between D and E hydrolysates of both 7S and 11S fractions, indicating sequential and simultaneous actions of trypsin and α-chymotrypsin had no differences on ACE inhibitory capacity. The IC$_{50}$ value of 11S-E was about 85% smaller than that of 7S-C, indicating more potent of the 11S digest as compared to 7S protein digest. The differences in the IC$_{50}$ values observed among 7S and 11S hydrolysates can be due to the differences in their structures, amino acid composition, and processing properties [27]. Furthermore, Gibbs et al. [6] found the biologically active peptides were mostly derived from glycinin (11S). The peptide YVVFK, which was resulted after the hydrolysis of the 11S protein, had been identified as a strong ACE inhibitor [4].

Table 2. Angiotensin convert enzyme (ACE) inhibitory capacity of the 7S and 11S hydrolysates.

Samples	IC$_{50}$ (mg/mL)	Samples	IC$_{50}$ (mg/mL)
7S-A	5.20 ± 0.13 [a,A]	11S-A	2.41 ± 0.38 [a,B]
7S-B	3.55 ± 0.03 [b,A]	11S-B	1.84 ± 0.06 [b,B]
7S-C	0.99 ± 0.01 [c,A]	11S-C	0.40 ± 0.04 [c,B]
7S-D	1.20 ± 0.01 [c,A]	11S-D	0.21 ± 0.02 [c,d,B]
7S-E	1.04 ± 0.01 [c,A]	11S-E	0.15 ± 0.07 [d,B]

Results are expressed as the mean ± standard deviation (on dried basis). Values followed by different superscript lowercase letters in the same column differ significantly ($p < 0.05$). Values followed by different superscript uppercase letters in the same row differ significantly ($p < 0.05$). IC$_{50}$: concentration to inhibit ACE by 50%. 7S and 11S: A—no hydrolyzed; B—hydrolyzed by pepsin for 2 h; C—hydrolyzed by pepsin for 2 h + trypsin for 2 h; D—hydrolyzed by pepsin for 2 h + trypsin for 2h + α-chymotrypsin for 2 h; and E—hydrolyzed by pepsin for 2 h + trypsin + α-chymotrypsin for 2 h.

Hydrolysis of 7S and 11S proteins by different proteases resulted in distinct peptide MW distribution for each fraction (Figure 2b,c). Most of peptides (>80%) present in protein hydrolysates exhibited MW below 2 kDa. Similar results were found in soy protein isolate (SPI) and soy protein hydrolysate (SPH) after simulated digestion [7]. In the non-hydrolyzed fractions (A), peptides of 4.5–10 kDa made up 77% and 55% of 7S-A and 11S-A, respectively. With pepsin, these peptides reduced to less than 10% and were unchanged with the addition of other enzymes for both fractions ($p > 0.05$). The percentage of peptides between 2 and 4.5 kDa was constant in all 7S and 11S hydrolysates (B, C, D, and E).

Peptides between 1 and 2 kDa were not found in the non-hydrolyzed 7S-A, but higher percentages were observed in the 7S-B and 7S-C, and they decreased in the 7S-D and 7S-E. While in the 11S protein, the relative content of theses peptides increased 2.9 fold in the 11S-B and unchanged in the 11S-C and 11S-D. The percentage of peptides of MW ranging from 0.5 to1 kDa increased until 7S-C and remained unaltered in the 7S-D and 7S-E hydrolysates. However, the relative content of these peptides remained unchanged with hydrolysis of 11S fraction ($p > 0.05$). No differences were observed in the percentage of peptides ranging from 0.1 to 0.5 kDa between 7S-A and 7S-B, but it increased in the 7S-C and reduced in the 7S-D and 7S-E. In the 11S-A, the relative content of these peptides was 21%, decreasing to 12% in the 11S-B and increasing in the 11S-D and 11S-E. The amount of the substances with MW < 0.1 kDa was about 10% in the 7S-A and was above 35% in the 7S-B, D, and E. The highest percentages of compounds <0.1 kDa in the 11S hydrolysates found in the 11S-B and 11S-C, were 31% and 28%, respectively, and 11S-E had lower percentage than 11S-D.

The difference in the peptide MW distributions among 7S and 11S hydrolysates probably was due to the diversity of generated peptides from each fraction. Glycinin (11S) is a hexamer with a molecular mass of 320–380 kDa, and each monomer subunit involves one basic and one acidic polypeptide linked via disulfide bond [3]. Whereas, β-conglycinin (7S) is a trimeric glycoprotein containing approximately 4% of carbohydrate with a molecular mass of 180 kDa consisting of three subunits associated by hydrophobic and hydrogen bonding [28]. Gibbs et al. [6] found glycinin was the precursor for 95% of the peptides isolated in their experiments, while the β-conglycinin was found to be more resistant to proteolytic attack.

3.4. Effect of Molecular Weight of Peptides on Angiotensin Converting Enzyme (ACE) Inhibitory Activity

The principal component analysis (PCA) analysis was performed between peptide MW and IC$_{50}$ values from simulated in vitro digested soy foods and 7S and 11S hydrolysates (Figure 3a,b). The projection on the factorial plane (FP1 × FP2) of peptide MW and IC$_{50}$ values is shown in the Figure 3a, and the projection of the soy foods and 7S and 11S hydrolysates is shown in the Figure 3b. The percentages of variance among the samples were 83% and 9% for PCA axes 1 and 2, respectively. The vectors of FP1 that were close indicated that the variables were positively correlated with each other and, therefore, a positive correlation was observed between the peptides of different MW, except that the substances smaller than 0.1 kDa had a lower correlation with others. The vectors that form an angle close to 180° indicated a negative correlation and, thus, peptides between 1 and 2 kDa were those that most contributed to the reduction of the IC$_{50}$ values, followed by the peptides between 2 and 4.5 kDa. Substances smaller than 0.1 kDa showed a weak negative correlation, indicating a smaller contribution to reducing the IC$_{50}$ value (Figure 3a). This result was consistent with reports by Puchalska, García, and Marina [10], who found the most potent ACE inhibitory activity were observed in peptides with MW below 3 kDa. Although studied intensively, the structure–function relationship between the ACE inhibitory activity and peptides remained unclear. Moreover, the number of amino acids in the composition of potentially antihypertensive peptides was not known. It might vary from two to several amino acids. Wu, Aluko, and Nakai [29,30] studied models for ACE-inhibitory peptides through computational analysis, and indicated that for dipeptides, amino acid residues with a large bulk chain as well as hydrophobic side chains are preferred such as phenylalanine, tyrosine, and tryptophan. The structure of the carboxyl terminal of a dipeptide is more relevant to the potency of ACE inhibitory activity than N-terminal. For tripeptides, the most favorable structure is to include an aromatic amino acid residue in the C-terminal, and the hydrophobic amino acids such as leucine, tryptophan in the N-terminal. Besides, the tetrapeptide residues from the C-terminal end determined the potency of peptides that contained 4 to 10 amino acid residues.

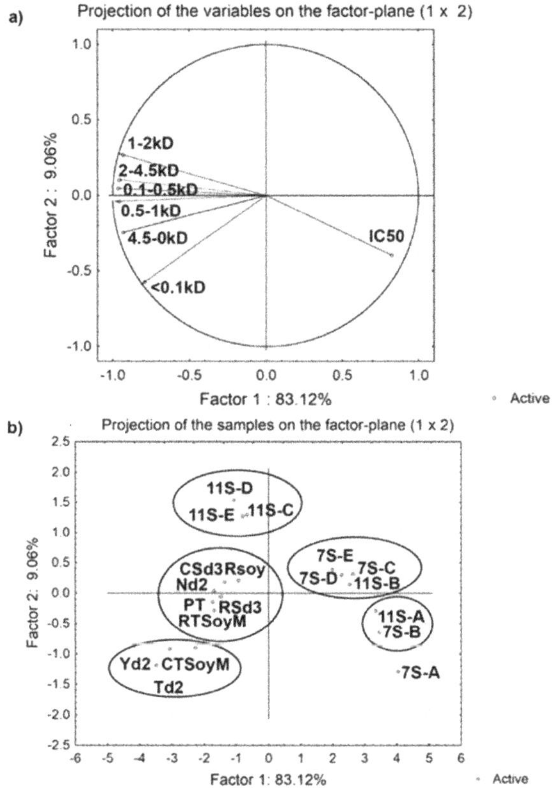

Figure 3. Scatterplots of the principal component analysis (PCA): (**a**) representing 8 variables and (**b**) representing 9 soybean foods, 5 hydrolysates of 7S and 5 hydrolysates of 11S. RSoy: raw soybean; RTSoyM: raw traditional soymilk; CTSoyM: cooked traditional soymilkPT: pressed tofu; RSd3: raw sprout germinated for 3 days; CSd3: cooked sprout germinated for 3 days; Nd2: natto stored for 2 days; Td2: tempeh stored for 2 days; and Yd2: yogurt stored for 2 days. 7S and 11S: A—no hydrolyzed; B—hydrolyzed by pepsin for 2 h; C—hydrolyzed by pepsin for 2 h + trypsin for 2 h; D—hydrolyzed by pepsin for 2 h + trypsin for 2 h + α-chymotrypsin for 2 h; and E—hydrolyzed by pepsin for 2 h + trypsin and α-chymotrypsin for 2 h.

Thus, additional studies on the identification of the active peptides of each type of soy foods should be performed in the future. In addition, it is necessary to clarify how the processing used to obtain soy foods affects the formation/degradation of bioactive peptides during the digestion through the human digestive tract.

Principal component analysis was used to group the samples according to the degree of similarity (Figure 3b), and the results showed that 7S-A, 7S-B, and 11S-A were located in the fourth quadrant and, therefore, were characterized by the higher IC_{50} values, and the 11S-A fraction was similar to 7S-B. The 11S-B was located in the first quadrant and grouped with 7S-C, D, and E. 11S-C, D, and E were grouped in the second quadrant, indicating that they were similar and with lower IC_{50} values. However, 11S-C, D, and E with low IC_{50} values did not show enough similarities to be clustered with the simulated in vitro digested soy foods. The digested soy foods Yd2, Td2, and CTSoyM were grouped in the third quadrant, whereas RTSoyM, RSd3, CSd3, PT, Nd2, and RSoy were clustered between quadrant 2 and 3. Therefore, results confirmed that the processing methods exerted an effect on protein hydrolysis during in vitro digestion and ACE inhibitory activity.

Considering the differences between digested soy foods and 7S and 11S hydrolysates, it was observed that peptides of MW between 1 and 4.5 kDa had a higher antihypertensive capacity. However, it was not possible to confirm that peptides in general were the only ones responsible for ACE inhibition activity, since results showed a lower IC_{50} value (53% less) of RTSoyM (a complex matrix) compared to 11S-E (isolated protein hydrolysate). For example, our previous research confirmed the higher ACE inhibitory activity of phenolics from various legume varieties [31]. Therefore, it is recommended to investigate the possibility of the presence of other types of compounds that have ACE inhibitory activity and if synergistic effects occur with other bioactive compounds. In addition, according to Capriotti et al. [4], extensive hydrolysis of soybean proteins during simulated gastrointestinal digestion generated a large number of peptides, some of which established biological activity. However, one should take into account that the in vivo digestion is a more complicated process than the in vitro model process, and the enzymes produced by the gut bacteria should also be considered.

4. Conclusions

The angiotensin converting enzyme (ACE) inhibitory properties of digested fermented and non-fermented soy and of hydrolysates of 7S and 11S fractions were reported for the first time in this study. Processing to manufacture soymilk, tofu, natto, tempeh, and soy yogurt improved the ACE inhibition after in vitro simulated gastrointestinal digestion. Raw and cooked traditional soymilk showed the highest anti-hypertensive activity. Among fermented soy foods, natto and soy yogurt had higher ACE inhibitory activity than tempeh. The germination time for 3 days was the most optimal to obtain sprouts with better anti-hypertensive activity. Hydrolysis of 7S and 11S fractions using pepsin followed by trypsin was enough for the released peptides with good ACE inhibitory capacity, and 11S hydrolysates were more powerful than 7S hydrolysates in ACE inhibition. Peptides between 1 and 4.5 kDa showed the highest ACE inhibitory activity. Further analysis is required for identification of peptides from digested soy foods with anti-hypertensive potential.

Author Contributions: Conceptualization, C.L.H. and S.K.C.C.; methodology, C.L.H., S.K.C.C., and Y.Z.; formal analysis, C.L.H.; investigation, C.L.H., J.X., and S.K.; resources, S.K.C.C.; writing—original draft preparation, C.L.H. and E.I.I.; writing—review and editing, C.L.H., S.K.C.C., E.I.I., Y.Z., J.X., and S.K.; supervision, S.K.C.C. All authors have read and agreed to the published version of the manuscript.

Funding: USDA-ARS SCA no. 58-6402-2729 contributed funding for this CRIS project no. MIS 501170.

Acknowledgments: Cíntia L. Handa. would like to thank Coordenação de Aperfeiçoamento de Pessoal de Nível Superior (CAPES) for exchange scholarship. Elza I. Ida. is a Conselho Nacional de Desenvolvimento Científico e Tecnológico (CNPq) Research Fellow. This work is a contribution of Mississippi Agricultural and Forestry Experiment Station, Mississippi State University.

References

1. Merai, R.; Siegel, C.; Rakotz, M.; Basch, P.; Wright, J.; Wong, B.; Thorpe, P.; Thorpe, P. CDC grand rounds: A public health approach to detect and control hypertension. *MMWR Morb. Mortal. Wkly. Rep.* **2016**, *65*, 1261–1264. [CrossRef] [PubMed]
2. Fitzgerald, R.J.; Murray, B.A. Bioactive peptides and lactic fermentations. *Int. J. Dairy Technol.* **2006**, *59*, 118–125. [CrossRef]
3. Fukushima, D. Recent progress in research and technology on soybeans. *Food Sci. Technol. Res.* **2001**, *7*, 8–16. [CrossRef]

4.   Capriotti, A.; Caruso, G.; Cavaliere, C.; Samperi, R. Identification of potential bioactive generated by imulated gastrointestinal digestion of soybean seeds and soy milk proteins. *J. Food Compos. Anal.* **2015**, *44*, 205–213. [CrossRef]

5.   Gu, Y.; Wu, J. LC-MS/MS Coupled with QSAR modeling in characterizing of angiotensin I-converting enzyme inhibitory peptides from soybean proteins. *Food Chem.* **2013**, *141*, 2682–2690. [CrossRef] [PubMed]

6.   Gibbs, B.F.; Zougman, A.; Masse, R.; Mulligan, C. Production and characterization of bioactive peptides from soy hydrolysate and soy-fermented food. *Food Res. Int.* **2004**, *37*, 123–131. [CrossRef]

7.   Coscueta, E.R.; Campos, D.A.; Osório, H.; Nerli, B.B.; Pintado, M. Enzymatic soy protein hydrolysis: A tool for biofunctional food ingredient production. *Food Chem. X* **2019**, *1*, 100006. [CrossRef]

8.   Li, F.; Ohnishi-Kameyama, M.; Takahashi, Y.; Yamaki, K. Angiotensin I-converting enzyme inhibitory activities of chinese fermented soypaste and estimation of the inhibitory substances. *J. Funct. Foods* **2013**, *5*, 1991–1995. [CrossRef]

9.   Alauddin, M.; Shirakawa, H.; Hiwatashi, K.; Shimakage, A.; Takahashi, S.; Shinbo, M.; Komai, M. Processed soymilk effectively ameliorates blood pressure elevation in spontaneously hypertensive rats. *J. Funct. Foods* **2015**, *14*, 126–132. [CrossRef]

10.   Puchalska, P.; Concepción García, M.; Luisa Marina, M. Identification of native angiotensin-I converting enzyme inhibitory peptides in commercial soybean based infant formulas using HPLC-Q-ToF-MS. *Food Chem.* **2014**, *157*, 62–69. [CrossRef]

11.   Hernández-Ledesma, B.; García-Nebot, M.J.; Fernández-Tomé, S.; Amigo, L.; Recio, I. Dairy protein hydrolysates: Peptides for health benefits. *Int. Dairy J.* **2014**, *38*, 82–100. [CrossRef]

12.   Zhang, Y.; Guo, S.; Liu, Z.; Chang, S.K.C. Off-flavor related volatiles in soymilk as affected by soybean variety, grinding, and heat-processing methods. *J. Agric. Food Chem.* **2012**, *60*, 7457–7462. [CrossRef] [PubMed]

13.   Meng, S.; Chang, S.; Gillen, A.M.; Zhang, Y. Protein and quality analyses of accessions from the USDA soybean germplasm collection for tofu production. *Food Chem.* **2016**, *213*, 31–39. [CrossRef] [PubMed]

14.   Wei, Q.; Wolf-Hall, C.; Chang, K.C. Natto characteristics as affected by steaming time, bacillus strain, and fermentation time. *J. Food Sci.* **2001**, *66*, 167–173. [CrossRef]

15.   Kumari, S.; Chang, S.K.C. Effect of cooking on isoflavones, phenolic acids, and antioxidant activity in sprouts of prosoy soybean (glycine max). *J. Food Sci.* **2016**, *81*, C1679–C1691. [CrossRef]

16.   Liu, C.; Wang, H.; Cui, Z.; He, X.; Wang, X.; Zeng, X.; Ma, H. Optimization of extraction and isolation for 11S and 7S globulins of soybean seed storage protein. *Food Chem.* **2007**, *102*, 1310–1316. [CrossRef]

17.   Cushman, D.W.; Cheung, H.S. Spectophotometric assay and properties of the angiotensin I-converting enzyme of rabbit lung. *Biochem. Pharmacol.* **1971**, *20*, 1637–1648. [CrossRef]

18.   You, S.J.; Udenigwe, C.C.; Aluko, R.E.; Wu, J. Multifunctional peptides from egg white lysozyme. *Food Res. Int.* **2010**, *43*, 848–855. [CrossRef]

19.   Smith, P.K.; Krohn, R.I.; Hermanson, G.T.; Mallia, A.K.; Gartner, F.H.; Provenzano, M.D.; Fujimoto, E.K.; Goeke, N.M.; Olson, B.J.; Klenk, D. Measurement of protein using bicinchoninic acid. *Anal. Chem.* **1985**, *150*, 76–85. [CrossRef]

20.   Lourenço da Costa, E.; Antonio da Rocha Gontijo, J.; Netto, F.M. Effect of heat and enzymatic treatment on the antihypertensive activity of whey protein hydrolysates. *Int. Dairy J.* **2007**, *17*, 632–640. [CrossRef]

21.   Zakharov, A.; Carchilan, M.; Stepurina, T.; Rotari, V.; Wilson, K.; Vaintraub, I. A comparative study of the role of the major proteinases of germinated common bean (*Phaseolus vulgaris* L.) and soybean (*Glycine max* (L.) merrill) seeds in the degradation of their storage proteins. *J. Exp. Bot.* **2004**, *55*, 2241–2249. [CrossRef] [PubMed]

22.   Yang, M.; Li, L. Physicochemical, textural and sensory characteristics of probiotic soy yogurt prepared from germinated soybean. *Food Technol. Biotechnol.* **2010**, *9862*, 490–496.

23.   Kirk, T.K.; Gifford, O.; Drive, P.; Farrell, R.L. Enzymatic "combustion": The microbial degradation of lignin. *Microbiology* **1987**, 465–505. [CrossRef] [PubMed]

24.   Barbana, C.; Boye, J.I. Angiotensin I-converting enzyme inhibitory activity of chickpea and pea protein hydrolysates. *Food Res. Int.* **2010**, *43*, 1642–1649. [CrossRef]

25.   Lahogue, V.; Réhel, K.; Taupin, L.; Haras, D.; Allaume, P. A HPLC-UV method for the determination of angiotensin I-converting enzyme (ACE) inhibitory activity. *Food Chem.* **2010**, *118*, 870–875. [CrossRef]

26.   Hayes, M.; Tiwari, B.K. Bioactive carbohydrates and peptides in foods: An overview of sources, downstream processing steps and associated bioactivities. *Int. J. Mol. Sci.* **2015**, *16*, 22485–22508. [CrossRef]

27. Nik, A.M.; Tosh, S.M.; Poysa, V.; Woodrow, L.; Corredig, M. Protein recovery in soymilk and various soluble fractions as a function of genotype differences, changes during heating, and homogenization. *J. Agric. Food Chem.* **2008**, *56*, 10893–10900. [CrossRef]

28. Thanh, V.H.; Shibasaki, K. Major proteins of soybean seeds. subunit structure of beta.-conglycinin. *J. Agric. Food Chem.* **1978**, *26*, 692–695. [CrossRef]

29. Wu, J.; Aluko, R.E.; Nakai, S. Structural requirements of angiotensin I-converting enzyme inhibitory peptides: Quantitative structure—Activity relationship study of Di- and tripeptides. *J. Agric. Food Chem.* **2006**, *54*, 732–738. [CrossRef]

30. Wu, J.; Aluko, R.E.; Nakai, S. Structural requirements of angiotensin I-converting enzyme inhibitory peptides: Quantitative structure-activity relationship modeling of peptides containing 4-10 amino acid residues. *QSAR Comb. Sci.* **2006**, *25*, 873–880. [CrossRef]

31. Zhang, Y.; Pechan, T.; Chang, S.K.C. Antioxidant and angiotensin-I converting enzyme inhibitory activities of phenolic extracts and fractions derived from three phenolic-rich legume varieties. *J. Funct. Foods* **2018**, *42*, 289–297. [CrossRef] [PubMed]

Article

Characterization and Demulsification of the Oil-Rich Emulsion from the Aqueous Extraction Process of Almond Flour

Fernanda F. G. Dias [1], **Neiva M. de Almeida** [2], **Thaiza S. P. de Souza** [1], **Ameer Y. Taha** [1] and **Juliana M. L. N. de Moura Bell** [1,3,*]

[1] Department of Food Science and Technology, University of California, Davis, One Shields Avenue, Davis, CA 95616, USA; ffgdias@ucdavis.edu (F.F.G.D.); thaizasps@gmail.com (T.S.P.d.S.); ataha@ucdavis.edu (A.Y.T.)

[2] Departamento de Gestão e Tecnologia Agroindustrial, Universidade Federal da Paraíba, Bananeiras, PB 58000-000, Brazil; neiva@cchsa.ufpb.br

[3] Biological and Agricultural Engineering, University of California, Davis, One Shields Avenue, Davis, CA 95616, USA

[*] Correspondence: jdemourabell@ucdavis.edu; Tel.: +1-530-752-5007

Received: 22 August 2020; Accepted: 27 September 2020; Published: 1 October 2020

Abstract: The aqueous extraction process (AEP) allows the concurrent extraction of oil and protein from almond flour without the use of harsh solvents. However, the majority of the oil extracted in the AEP is present in an emulsion that needs to be demulsified for subsequent industrial utilization. The effects of scaling-up the AEP of almond flour from 0.7 to 7 L and the efficiency of enzymatic and chemical approaches to demulsify the cream were evaluated. The AEP was carried out at pH 9.0, solids-to-liquid ratio of 1:10, and constant stirring of 120 rpm at 50 °C. Oil extraction yields of 61.9% and protein extraction yields of 66.6% were achieved. At optimum conditions, enzymatic and chemical demulsification strategies led to a sevenfold increase (from 8 to 66%) in the oil recovery compared with the control. However, enzymatic demulsification resulted in significant changes in the physicochemical properties of the cream protein and faster demulsification (29% reduction in the incubation time and a small reduction in the demulsification temperature from 55 to 50 °C) compared with the chemical approach. Reduced cream stability after enzymatic demulsification could be attributed to the hydrolysis of the amandin α-unit and reduced protein hydrophobicity. Moreover, the fatty acid composition of the AEP oil obtained from both demulsification strategies was similar to the hexane extracted oil.

Keywords: aqueous extraction process; extraction yields; cream demulsification; oil recovery; almond flour

---

1. Introduction

Almond (*Prunus dulcis* L.) is the most widely consumed tree nut in the United States, with the state of California being responsible for 100% of the U.S. almond supply and 80% of the world's almond production [1]. The consumption of almonds has been linked to nutritional and health benefits arising from its composition [2,3]. Almonds are rich in oil, protein, and carbohydrates, as well as micronutrients such as vitamin E, manganese, magnesium, copper, phosphorus, fiber, and riboflavin [4]. The lipid content in almonds is around 55% and consists mainly of polyunsaturated oleic and linoleic fatty acids [4,5]. Almonds also contain a high protein content (27%), which contributes significantly to nutrition, flavor, and other important functional properties when used as the main ingredient in several food products [6].

Almond proteins possess unique functional properties that make them a desirable ingredient for use in several food applications, especially plant-based beverages. Similarly, almond oil is also known for its health-promoting benefits (anti-inflammatory and anti-hepatotoxic) [7] and less harsh and environmentally friendly extraction methods are needed. However, the isolation of almond proteins for subsequent food/nutraceutical applications requires prior oil removal and the development of environmentally friendly strategies able to preserve the functional properties of the extracted protein. Oil extraction is commonly achieved by solvent extraction and/or mechanical pressing, the choice of which depends on the composition of the starting material. Pressing is usually the method of choice to produce specialty oils (i.e., tree nuts), however, it might generate a high oil protein-meal that needs to be subjected to subsequent solvent extraction to improve oil removal [8,9]. However, the use of hazardous solvents (i.e., hexane) has raised several environmental, safety, and consumer concerns [10,11]. In response to these concerns, research has been undertaken to develop alternative environmentally friendly strategies [12–14] to fractionate oil and protein from many oil-bearing materials. The aqueous extraction process (AEP) is a promising alternative for the fractionation of the almond constituents (i.e., protein, lipids, and carbohydrates) for subsequent use in the development of new food products and nutraceuticals. This water-based technology allows the simultaneous extraction of oil and proteins without the use of solvents, with potential value-added products [10,15–18]. Although this technique has been more extensively evaluated for soybeans [10,16,19], peanuts [20,21], linseed [22], flaxseed [23], olive [24,25], and corn [26,27], its use for the fractionation and production of almond ingredients for food and nutraceutical applications is still in its early stages [28–30].

The AEP of almond flour and almond cake (the protein-rich meal generated by the mechanical expression of the almond oil) achieved oil extraction yields from 50 to 70% at laboratory scale [28,30]. However, only a reduced fraction of the almond oil was extracted as free oil (1 to 3%) [28,30]. As a matter of fact, 25–27% of the almond cake oil [30] and 40–64% of the almond flour oil [28], although extracted from their respective starting material, ended up entrapped in the cream emulsion. Therefore, the development of viable methods to demulsify the cream emulsion, a necessary step to increase the recovery of the extracted oil, is essential to maximize the feasibility of the AEP [31–33].

The stability of the cream emulsion depends on the molecular and chemical properties of the emulsifier at the interface and on the process conditions [33,34]. Enzymatic and physical treatments have been exploited to reduce the cream stability and increase free oil release from the AEP and enzyme-assisted aqueous extraction processing (EAEP) of soybeans [10,33,35,36], peanut [20], yellow mustard [37], sunflower [38], and almond cake [29]. However, the effects of different demulsification strategies on the release of free oil from the almond cream and resulting physicochemical changes in the cream remain a challenge to the development of a structural-based approach to improve oil recovery and quality.

The objectives of this study were to (i) scale up the aqueous extraction process of almond flour from 0.07 to 0.7 kg of almond flour using a 1:10 solids-to-liquid ratio (SLR) (~7 L of slurry) to produce adequate quantities of cream for the demulsification and characterization experiments, (ii) assess the distribution of the extracted oil and protein among the fractions generated (skim, cream, and insoluble), (iii) evaluate the resistance of the cream against enzymatic and chemical demulsification strategies, (iv) determine the effects of chemical and enzymatic demulsification strategies in the physicochemical properties of the cream proteins (peptide profile and surface hydrophobicity), and (v) determine the fatty acid profile of the AEP almond oil. With the goal of maximizing free oil recovery, sequential optimization strategies using response surface methodology were employed to destabilize the cream produced by the AEP.

2. Materials and Methods

2.1. Material

The almond flour used in this study (ultra-fine granulometry) was provided by Blue Diamond Growers (Sacramento, CA, USA). Whole almonds (*Prunus dulcis*) were grounded and sieved in a

#12 mesh sieve (minimum recovery of 85%). The almond flour composition was 43% oil, 22% protein, and 5% moisture (determined according to Section 2.6).

2.2. Processing Scale-Up of the Aqueous Extraction Process (AEP) of Oil and Protein from Almond Flour

Almond flour (0.7 kg) was mixed with water to achieve a 1:10 solids-to-liquid ratio (SLR) in a 10 L jacketed glass reactor model CG-1965-610M (Chemglass Life Sciences LLC, Vineland, NJ, USA). The slurry pH was adjusted and kept at 9.0 for 60 min. The slurry pH was constantly monitored with a pH meter and it was maintained at pH 9.0 with the addition of 2 M NaOH or 2 MHCl. Extractions were performed at 50 °C under constant agitation at 120 rpm. Extraction conditions were previously optimized by Almeida et al. [28] at laboratory-scale. After the extraction, the slurry was centrifuged at 3000× g for 30 min to separate the insoluble fraction (Figure 1) from the liquid phase (cream and skim). The liquid phase was cooled to 4 °C and subsequently fractionated into (i) skim–protein-rich fraction and (ii) cream–oil-rich fraction. The separation was performed based on the density difference among the fractions; the skim was the lower phase and the cream was the upper phase The fractionation of cream and skim was performed by opening the chemical reactor valve and letting the skim fraction flow until the cream fraction reached the valve. The cream fraction was subsequently collected in a separate container. The weight of both fractions was recorded for mass balance calculations. The extraction process was performed in triplicate.

The AEP fractions (skim, cream, and insoluble) and the starting material (almond flour) were evaluated for oil and protein content. Total oil extraction (TOE) yield, oil distribution in the fractions, total protein extraction (TPE) yield, and protein distribution in the fractions were determined according to Equations (1)–(4), respectively [28].

$$\text{TOE (\%)} = \left[100 - \left(\frac{\text{Oil (g) in the insoluble fraction}}{\text{Oil (g) in the almondflour}}\right)\right] \times 100 \tag{1}$$

$$\text{Oil distribution in each fraction (\%)} = \left(\frac{\text{Oil (g) in each fraction*}}{\text{Oil (g) in the almond flour}}\right) \times 100 \tag{2}$$

$$\text{TPE (\%)} = \left[100 - \left(\frac{\text{Protein (g) in the insoluble fraction}}{\text{Protein (g) in the almond flour}}\right)\right] \times 100 \tag{3}$$

$$\text{Protein distribution in each fraction (\%)} = \left(\frac{\text{Protein (g) in each fraction*}}{\text{Protein (g) in the almond flour}}\right) \times 100 \tag{4}$$

* with fraction corresponding to either cream, skim, or insoluble.

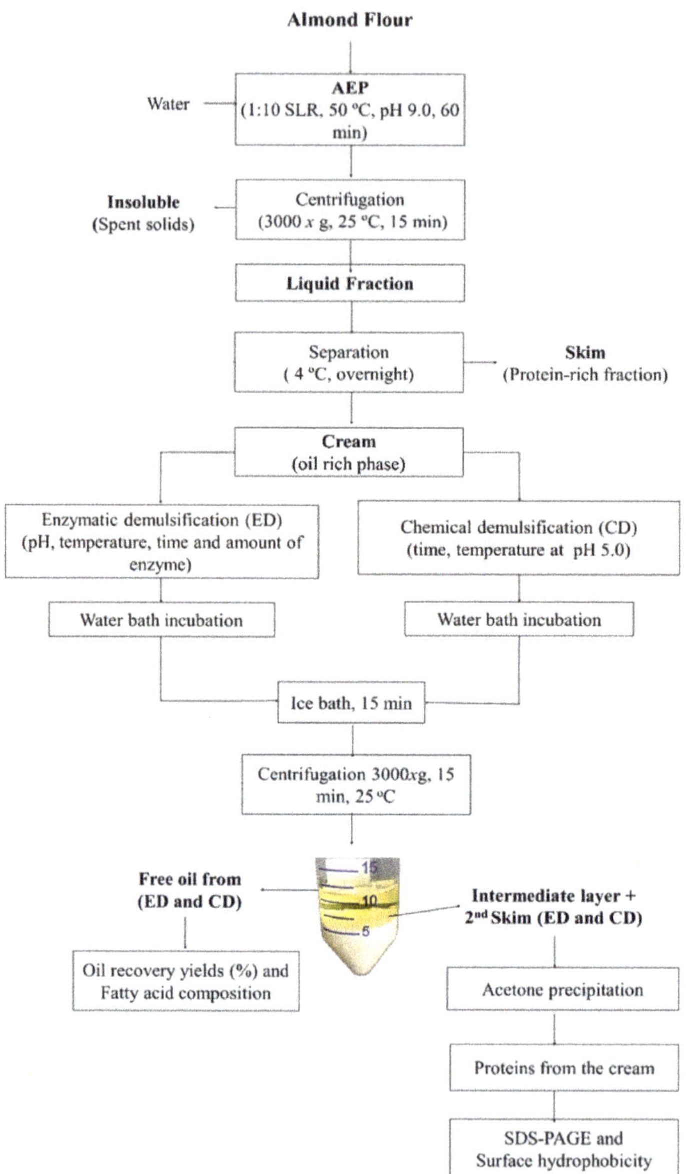

Figure 1. Process flow diagram for the aqueous extraction process (AEP) of almond flour and demulsification (ED: enzymatic demulsification and CD: chemical demulsification) strategies to recover the oil from the cream emulsion. SLR, solids-to-liquid ratio; SDS-PAGE, sodium dodecyl sulfate-polyacrylamide gel electrophoresis.

2.3. Recovering the Extracted Oil: Enzymatic (ED) and Chemical Demulsification (CD) of the Almond Cream

Considering that the majority of the oil extracted by the AEP is entrapped in the cream [18,31,33], the development of strategies to demulsify the cream is necessary to free the entrapped oil for subsequent utilization. Understanding the effects of processing variables (i.e., pH, time, temperature, amount of enzyme) on the release and quality of the extracted oil is a key step to maximize the feasibility of the AEP.

Overall, cream demulsification was performed using 10 g of well-mixed cream (cream + free oil) placed in a 30 mL beaker. For the enzymatic demulsification, the cream pH was adjusted to different pH values (Table 1, fractional factorial design matrix) using a 2 N NaOH solution before the addition of the Neutral Protease 2 million (NP2M) (BIO-CAT, Virginia, NY, USA). Enzyme selection was based on the ability of this enzyme to extract a higher amount of almond oil in its free form in laboratory scale experiments [28]. No enzyme was added during the chemical demulsification and control sample after pH adjustment. Cream demulsification was performed under constant stirring (120 rpm) and controlled temperature in a water bath with a multipoint inductive-drive stirrer (ThermoScientific Variomag, Thermo Scientific, Daytona Beach, FL, USA). After the demulsification treatment, samples were transferred to a 50 mL centrifuge tube and centrifuged at 3000× *g* for 15 min at 25 °C. Three distinct layers were obtained (free oil, an intermediate layer, and a water phase referred to as second skim) (Figure 1). Most of the free oil was collected using a Pasteur pipette and the remaining free oil was extracted by rinsing the cream twice with hexane (2 mL in total), which was subsequently evaporated under nitrogen flux, as described by Lamsal and Johnson 2007. The free oil yield (%) was calculated as follows (5) [12]:

$$\text{Demulsification yield } (\%) = \frac{\text{free oil } (\text{g}) \times 100}{[\text{cream } (\text{g}) \times \text{oil content } (\%)] / 100\%} \tag{5}$$

2.3.1. Enzymatic Demulsification (ED): Tailoring Enzyme Use to Maximize the Cream Demulsification Efficiency

A sequential strategy based on the use of a fractional factorial design (FFD) followed by a central composite rotatable design (CCRD) was used to optimize the enzymatic demulsification of the cream emulsion. The effects of pH (6.0–9.0), amount of enzyme (0.5–1.0%), time (30–90 min), and temperature (50–65 °C) on the destabilization of the cream emulsion were evaluated using a FFD. The fractional factorial design matrix is described in Table 1. The levels of the variables used in the experimental design were selected based on preliminary experiments, optimum conditions of enzyme activity, and existing literature data [15,32]. Cream demulsification yield (%) was the dependent variable, indicating the amount of oil recovered from the cream emulsion.

According to the FFD experiments, pH and amount of enzyme significantly affected the stability of the cream emulsion. On the basis of the significance and type of effect (i.e., positive or negative) of pH and amount of enzyme observed in the FFD, a CCRD with three central points and four axial points (total of 11 runs) was developed to maximize the cream demulsification yield. The variables with negative effects had their values shifted to lower levels, and the variables with positive effects had their values shifted to higher levels in the CCRD. Therefore, the effects of pH (6.6–9.4) and amount of enzyme (0.1–1.51%) on the demulsification of the cream were evaluated. Demulsification yields were calculated as described in Equation (5) and the second-order equation used for the model is represented by Equation (6):

$$Y = \beta_0 + \sum_{i=1}^{n} \beta_i x_i + \sum_{i=1}^{n-1} \sum_{j=i+1}^{n} \beta_{ij} x_i x_j \tag{6}$$

where Y is the estimated response, i and j are values from 1 to the number of variables (n), β_0 is the intercept term, β_i are the linear coefficients, β_{ij} are quadratic coefficients, and x_i and x_j are coded independent variables. The significance of the model was tested by analysis of variance (ANOVA). Best experimental conditions determined by the CCRD were subsequently validated in triplicate and compared with the predicted value generated by the regression model.

2.3.2. Chemical Demulsification (CD): Effects of Temperature and Incubation Time on the Cream Demulsification Efficiency

For the chemical demulsification, the pH of the cream was kept at 5.0, which is the almond protein isoelectric point [39]. The effects of incubation temperature (46.9–68.1 °C) and time (5.8–69.2 min) on the demulsification of the cream were evaluated. Cream demulsification yield (%) was calculated as described in Equation (5). The best experimental condition identified by the CCRD was validated in triplicate and compared with the predicted value generated by the regression model.

2.4. Effects of Demulsification Strategies on the Physicochemical Properties of the Cream Proteins

2.4.1. Low Molecular Weight (MW) Polypeptide Profile Characterization of the Cream Proteins by Sodium Dodecyl Sulfate-Polyacrylamide Gel Electrophoresis (SDS-PAGE)

The proteins in the intact cream (IC) (before demulsification) and in the chemically (CD) and enzymatically (ED) demulsified creams were isolated as described by Chabrand and Glatz [35]. Briefly, proteins in the cream oil/water interface were isolated by precipitation with ice-cold acetone in the proportion of 1:20 (sample/acetone), followed by incubation at −20 °C for 2 h and centrifugation at 12,000× g for 15 min at 4 °C. The protein pellets obtained were washed three times with cold acetone and air-dried at 25 °C. Air-dried proteins were mixed with Laemmli buffer containing β-mercaptoethanol and placed in a water bath at 95 °C for 5 min as described by Laemmli [40]. A total of 30 μg of protein was loaded per well onto a precast gradient acrylamide 4–20% gel (Criterion TM TGX Precast Gels, Bio Rad, Hercules, CA, USA). A 25 mM Tris buffer pH 8.3 containing 192 mM glycine and 0.1% of sodium dodecyl sulfate (Bio Rad, Hercules, CA, USA) was used as the running buffer at 200 V at room temperature for 60 min. A Precision Plus Protein™ Dual Color Standards SDS-PAGE standard (0–250 kDa) (Bio Rad, Hercules, CA, USA) was used as the molecular weight size marker. Relative quantification and polypeptide distribution were performed using a Gel Doc TM EZ Imager system and Image Lab software (Bio-Rad, Hercules, CA, USA).

2.4.2. Surface Hydrophobicity (H_0) of the Cream Proteins

The H_0 of the cream proteins was determined using the ANS (1-anilino-8-naphthalenesulfonate) fluorescence probe as described by Zhang et al. [41] with a few modifications. Protein samples from intact cream (before demulsification, IC) and after chemical (pH treated cream, CD) and enzymatic (enzymatically treated cream, ED) demulsification (prepared as described in item 2.4) were dispersed into a 100 mM PBS (phosphate-buffered saline solution) at pH 7.0 and centrifuged at 10,000× g for 20 min. The amount of protein in the supernatant was quantified by the fluorometric based method Qubit (Protein assay kit, ThermoFisher, Waltham, MA, USA) using the Qubit Fluorometer 4.0 (ThermoFisher, Waltham, MA, USA). The amount of protein was adjusted to concentrations ranging from 0.04 to 0.70 mg/mL and 1.25 μL of ANS solution (8.0 mM in 100 mM PBS, pH 7.0, solution) was added to 250 μL of sample solution in a 96-well plate. Fluorescence intensity was measured in a plate reader (SpectraMax M5, Molecular Devices, San Jose, CA, USA) at 390 and 470 nm of excitation and emission wavelength, respectively. Sample hydrophobicity was defined as the slope of the fluorescence intensity versus protein concentration curve and was calculated by linear regression analysis [31]. The H_0 value was calculated as the average of six measurements.

2.5. Fatty Acid Composition of the Almond Oil Recovered by Chemical and Enzymatic Demulsification Strategies and Solvent Extraction

Total fatty acids were measured by gas chromatography coupled to a flame ionization detector (GC-FID). An aliquot of 400 μL of toluene spiked internal standard (triheptadecanoic acid) was added to 10 mg of almond oil (oil recovered from chemical and enzymatic demulsification and solvent extraction), followed by the addition of 3 mL of 100% methanol and 600 μL of HCl/methanol (8:92 *v/v*). Sample derivatization was performed at 90 °C for 60 min followed by a cooling step of 5 min at room temperature. The fatty acid methyl esters (FAMEs) were extracted using 1 mL of hexane and 1 mL of ultrapure water and separating the hexane layer (containing the FAMEs) to a new centrifuge tube containing 450 μL of ultrapure water. Samples were centrifuged at 15,000× g for 1 min at 0 °C, and the top hexane layer was transferred to a new tube, dried under nitrogen, and reconstituted in 200 μL of hexane.

FAMEs were analyzed by a gas chromatograph (GC, Clarus 500, Perkin Elmer, Waltham, MA, USA) coupled with a flame ionization detector (FID) with helium as the carrier gas at 1.0 mL/min. The column used was a DB-FFAP (30 m × 0.25 mm id × 0.25 μm; Agilent, Santa Clara, CA, USA). The injector temperature was 240 °C with a split of 1:10 and injection volume of 1 μL. The oven program temperature was 80 °C for 2 min, increased to 180 °C at 10 °C/min, increased to 240 °C at 5 °C/min, and held at 240 °C for 13 min. The detector temperature was kept at 300 °C. A mix of 29 FAME standards was used to identify the different fatty acids (FAs) based on their retention times. For comparison purposes, hexane extracted almond oil was used as the control for the oil recovered by chemical and enzymatic demulsification strategies. Almond oil solvent extraction was performed using a Soxhlet apparatus for 5 h at 65–68 °C using hexane as the solvent, according to the AOAC official method for oil extraction [42]. The hexane was removed from the oil under nitrogen flux and the oil was stored at −80 °C until further analysis.

2.6. Proximate Analyses

Total oil content was determined by the Mojonnier method (AOAC Method 922.06 for solid samples, and AOAC Methods 995.19 and 989.05 for cream and skim fractions, respectively) [43]. Nitrogen content was determined using the Dumas method (Vario Max Cube, Elementar Americas, Ronkonkoma, NY, USA). Protein content (%) was calculated as total nitrogen (%) × 5.18 (conversion factor). Total solids (dry matter) was determined by weighing after drying the samples in a vacuum oven at 65 °C until constant weight (AACC Method 44–40) [44]. All analyses were conducted in duplicate for each extraction replicate ($n = 3$) and a mass balance was provided for all extracted compounds.

2.7. Statistical Analyses

Multivariate statistical analyses were performed for the FFD and CCRD using the Protimiza Experimental Design® Software (http://experimental-design.protimiza.com.br). GraphPad Prism (Version 7.0, La Jolla, CA, USA) was used for additional experiments. Replicates of each measurement were analyzed by analysis of variance (ANOVA) followed by Tukey with a level of significance set at $p < 0.05$.

3. Results and Discussion

3.1. Processing Scale-Up of the Aqueous Extraction Process of Oil and Protein from Almond Flour

To ensure the production of adequate quantities of cream (oil-rich fraction) for subsequent demulsification and compositional analyses, optimum extraction conditions identified at laboratory scale for the AEP (pH 9.0, 50 °C, 1:10 SLR, 60 min of extraction time) were scaled-up from 0.07 to 0.7 kg of almond flour, producing ~7 L of slurry. Total oil and protein extraction yields and their distribution in the fractions are shown in Figure 2A,B. Total oil and protein extraction yields of 61.9 and 66.6% were achieved at pilot scale, respectively. These results are in agreement with those obtained at laboratory scale (0.07 kg of almond flour), where oil and protein extraction yields of 69.8 and 64.8% were achieved, respectively [28]. Overall, the distribution of the extracted oil and protein in the skim and cream fractions was in agreement with that observed at lab-scale experiments. Low oil yield (<5%) was observed in the skim fraction, a desirable finding because the presence of oil in the skim might reduce the skim protein solubility, and there are no feasible methods to remove the oil from the skim fraction [12]. Although 61.9% of the almond flour oil was extracted, 93.5% of extracted oil was entrapped in the cream fraction when scaling up the extraction process (Figure 2A). The cream fraction contained approximately 1.5% of the almond protein.

Because proteins can behave as potent emulsifiers [20], the cream protein content and composition might influence the resistance of the cream against subsequent demulsification (i.e., the recovery of the oil from the emulsion) [45].

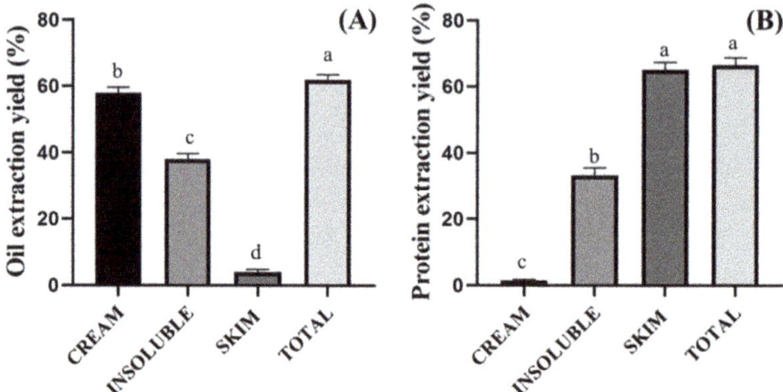

Figure 2. Extraction yields and distribution of oil (**A**) and protein (**B**) among the fractions generated (cream, insoluble, and skim) by the aqueous extraction process of almond flour. Different letters indicate a significant difference between extraction yields among the fractions within each compound (oil and protein) by the Tukey's test ($p < 0.05$).

Although tailoring extraction conditions to maximize overall extractability of oil and protein from almond flour is unquestionably the first step in the process development, it is important to understand the effects of extraction conditions on the recovery of the extracted oil (commonly entrapped in the cream emulsion) for subsequent applications (i.e., food, feed, and fuel).

3.2. Recovering the Extracted Oil: Enzymatic and Chemical Demulsification of the Cream Emulsion

In view of the fact that most of the extracted oil in the AEP is entrapped in the cream (oil-rich fraction), it becomes necessary to carefully consider the techniques employed to free the oil from the cream emulsion. Maximizing oil recovery and oil quality is key to process feasibility and will depend critically on the development of successful demulsification steps [18]. Because demulsification conditions (e.g., use of enzyme, pH, temperature, and incubation time) can impact the destabilization of the cream, we have employed different statistical approaches to optimize the enzymatic and chemical demulsification of the cream.

3.2.1. Enzymatic Demulsification: Tailoring Enzyme Use to Maximize Cream Demulsification Efficiency

The effects of pH, amount of enzyme, reaction time, and temperature were evaluated on the enzymatic demulsification of the cream by an FFD. Fractional factorial design matrix (2^{4-1}, with four independent variables and three repetitions in the central point) and the respective responses for each experiment are presented in Table 1. The independent variables were pH, temperature, reaction time, and amount of enzyme; the dependent variable was the cream demulsification yield. Higher free oil recoveries were observed in experiment #6 (55.6% at pH 9.0 and 50 °C), followed by experiments #4 (49.6%) and #9–10 (central points) (48.5 ± 1.2%), with free oil recovery varying from 18 to 55%. These results highlight the importance of demulsification parameters on the destabilization of the cream.

Table 1. Effects of pH, temperature, reaction time, and amount of enzyme on cream demulsification yield using a fractional factorial design.

Experiments	Experimental Parameters				Demulsification Yield (%)
	X_1 (pH)	X_2 (T, °C)	X_3 (t, min)	X_4 (E, %)	
1	−1 (6.0)	−1 (50.0)	−1 (30)	−1 (0.5)	18.6
2	1 (9.0)	−1 (50.0)	−1 (30)	1 (2.5)	20.4
3	−1 (6.0)	1 (65.0)	−1 (30)	1 (2.5)	17.8
4	1 (9.0)	1 (65.0)	−1 (30)	−1 (0.5)	49.6
5	−1 (6.0)	−1 (50.0)	1 (90)	1 (2.5)	19.2
6	1 (9.0)	−1 (50.0)	1 (90)	−1 (0.5)	55.6
7	−1 (6.0)	1 (65.0)	1 (90)	−1 (0.5)	24.3
8	1 (9.0)	1 (65.0)	1 (90)	1 (2.5)	31.2
9	0 (7.5)	0 (57.5)	0 (60)	0 (1.5)	47.4
10	0 (7.5)	0 (57.5)	0 (60)	0 (1.5)	49.8
11	0 (7.5)	0 (57.5)	0 (60)	0 (1.5)	48.4

pH is the pH of the cream, T is the temperature (°C), t is incubation time (min), and E is the amount of enzyme added in the cream (%, weight of enzyme/weight of the cream emulsion). Fractional factorial design matrix (2^{4-1}, with four independent variables and three repetitions in the central point).

The FFD results showed that the pH had a significant and positive effect (19.96) on the demulsification yield of the cream (Table 2), indicating that higher pH values favor oil release. The observed effect is likely because enzyme activity increases when pH increases from 6 to 8, but it is slightly reduced at higher pH values such as pH 9.0. The amount of enzyme used during the demulsification had a negative effect (−15.62) on demulsification yields, indicating that increasing the amount of enzyme from 0.5 to 2.5% would reduce free oil recovery. A similar trend, where increased oil recovery was observed up to certain enzyme concentration followed by steady or decreased free oil recovery, has been reported by Jiang et al. [21]. Such behavior could be attributed to extensive hydrolysis, which might increase emulsion formation [20]. Overall, enzyme concentration depends on the composition of the cream, enzyme cost, and quality of the extracted oil [18,29,46].

Within the range evaluated, incubation time (30–90 min) and temperature (50–65 °C) did not significantly affect the cream demulsification efficiency, indicating that the temperature values and the incubation times evaluated in this study were within the optimal range for the enzyme activity. These results indicate that the destabilization of the cream could be achieved in a shorter reaction time (30 min) and at a lower temperature (50 °C), which would reduce energy costs while better preserving the quality of the extracted oil. Overall, the FFD results indicate that enzymatic destabilization of the AEP cream can be achieved by the use of a reduced amount of enzyme (<2.5%) and higher pH (>6.0).

Table 2. Estimated effects of pH, temperature, incubation time, and amount of enzyme on the enzymatic demulsification yield of the cream emulsion.

Variables *	Effect	*p*-Value
pH (X_1)	19.96	0.02
Temperature (°C) (X_2)	1.52	0.80
Time (min) (X_3)	6.75	0.30
Enzyme (%) (*w/w*) (X_4)	−15.62	0.05

* $R^2 = 0.86$.

On the basis of the significance of the effects of the variables evaluated in the FFD, a CCRD was used to optimize the reaction pH and amount of enzyme to maximize free oil recovery. Time and temperature were fixed at their lowest levels (30 min and 50 °C) as both variables did not significantly affect demulsification yields. The values of the variables with significant effects (pH and amount of enzyme) were shifted to higher (from 6.0–9.0 to 6.6–9.4 for the pH) or lower values (from 0.5–2.5%

to 0.1–1.51% for the amount of enzyme) depending on the type of effect observed (i.e., positive or negative).

The CCRD showed that higher demulsification yields were obtained at the central points (experiments #5, 6, and 7: 63.1 ± 3.6%, pH 8 and 0.8% of enzyme) and the lowest yield (37.2%) was observed in experiment #8 (pH 6.6 and 0.8% of enzyme), where the lowest pH value was used (Table 3). The highest demulsification yield obtained in the CCDR was 13% higher than the maximum demulsification yield reported in the FFD, highlighting that the preliminary evaluation of the effects of the demulsification variables followed by the optimization of selected variables was a useful approach to further increase free oil recovery.

Table 3. Effects of pH and amount of enzyme on the enzymatic demulsification of the cream.

Experiments	Experimental Parameters		Demulsification Yield (%)
	X_1 (pH)	X_2 (E, %)	
1	−1 (7.0)	−1 (0.3)	50.0
2	1 (9.0)	−1 (0.3)	40.4
3	−1 (7.0)	1 (1.3)	39.0
4	1 (9.0)	1 (1.3)	42.6
5	0 (8.0)	0 (0.8)	60.2
6	0 (8.0)	0 (0.8)	62.1
7	0 (8.0)	0 (0.8)	67.1
8	$-\alpha$ * (6.6)	0 (0.8)	37.2
9	α (9.4)	0 (0.8)	46.2
10	0 (8.0)	$-\alpha$ (0.1)	42.9
11	0 (8.0)	α (1.5)	39.5

pH is the pH of the cream, and E is the enzyme amount added in the cream (%, weight of enzyme/weight of the cream emulsion). * Central composite rotatable design (2^2, with two independent variables, three repetitions in the central point and four axial points), $\alpha = 1.4142$.

The individual and interaction effects between pH and amount of enzyme were determined by multiple regression analysis of the experimental data (Table 4). Only regression coefficients statistically significant at $p < 0.05$ were used in the reparameterized models. The regression model for enzymatic demulsification ($Y_{ED}(\%) = 63.11 - 10.31X_1^2 - 10.57X_2^2$) and the response surface (Figure 3A) indicate that free oil release is influenced by the pH (X_1) and amount of enzyme (X_2). The reparametrized model was able to explain 86% of the total variation between the observed and predicted values. The regression was statistically significant ($F_{calculated}$ (25.4) > $F_{tabulated}$ (4.5)), whereas the F-test for lack of fit was not statistically significant ($F_{calculated}$ (1.6) < $F_{tablulated}$ (19.3)) at $p < 0.05$ (Table S1—Supplementary Material). According to the predictive model and response surface (Figure 3A), 63% free oil recovery can be achieved by enzymatic demulsification when using pH 8.0 and 0.8% (*w/w*) of enzyme.

Table 4. Analysis of variance (ANOVA) including models, R^2, and probability for the final reduced models for the chemical and enzymatic cream demulsification.

Response	Equation	F-Test	R^2	*p*-Value
ED	$Y_{ED}(\%) = 63.11 - 10.31X_1^2 - 10.57X_2^2$	5.64	0.8639	0.00034
CD	$Y_{CD}(\%) = 67.35 - 3.49X_1 - 9.34X_1^2 + 3.33X_2 - 7.10X_2^2$	5.14	0.9368	0.00096

ED: enzymatic demulsification; Y_{ED} is the dependent variable, enzymatic demulsification yield (%); CD: chemical demulsification; Y_{CD} is the dependent variable, chemical demulsification yield (%); X_1 and X_2 are the independent variables (pH, and amount of enzyme for the ED and temperature and time for the CD).

(A)   (B)

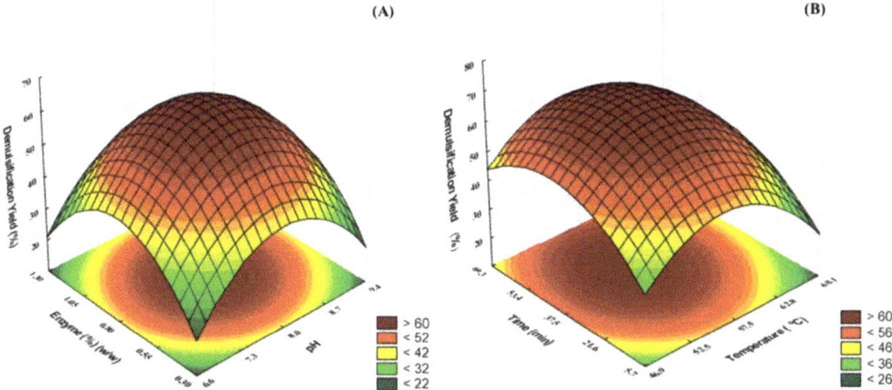

Figure 3. Response surface plot for the (**A**) enzymatic and (**B**) chemical demulsification of the AEP cream.

3.2.2. Chemical Demulsification: Effects of Temperature and Incubation Time on the Cream Demulsification Efficiency

A central composite rotatable design (CCRD) was used to optimize the incubation time (5.8 to 69.2 min) and the reaction temperature (46.9 to 68.1 °C) during the chemical demulsification of the cream. Table 5 shows the CCDR matrix and the respective responses to maximize chemical demulsification yields. The pH was kept at 5.0 [39] to favor protein precipitation, which could in turn minimize electrostatic repulsion between the oil droplets and increase free oil recovery. Maximum demulsification yields (67.3 ± 2.0%) were achieved in the central points (experiments #5, 6, and 7: 57.5 °C and 37.5 min), with the lowest oil recovery being observed at the axial point, experiment #10 (42.5%), where the shortest reaction time was used (5.8 min).

Table 5. Effects of reaction time and temperature on chemical demulsification of the cream.

Experiments	Experimental Parameters		Demulsification Yield (%)
	X_1 (T, °C)	X_2 (t, min)	
1	−1 (50.0)	−1 (15.0)	49.9
2	1 (65.5)	−1 (15.0)	47.7
3	−1 (50.0)	1 (60.0)	59.0
4	1 (65.5)	1 (60.0)	46.2
5	0 (57.5)	0 (37.5)	67.4
6	0 (57.5)	0 (37.5)	69.3
7	0 (57.5)	0 (37.5)	65.3
8	−α* (46.96)	0 (37.5)	53.7
9	α (68.1)	0 (37.5)	43.9
10	0 (57.5)	−α (5.8)	42.5
11	0 (57.5)	α (69.2)	57.9

T is the temperature and t is the incubation time. * Central composite rotatable design (2^2, with two independent variables, three repetitions in the central point and four axial points), $\alpha = 1.4142$.

The individual and interaction effects between time and temperature were determined by multiple regression analysis of the experimental data (Table 4). Regression coefficients not statistically significant at $p < 0.05$ were disabled in the reparametrized model to a significance level of 5%. The regression model for enzymatic demulsification ($Y_{CD}(\%) = 67.35 - 3.49X_1 - 9.34X_1^2 + 3.33X_2 - 7.10X_2^2$) and response surface (Figure 3B) show that free oil release is influenced by the linear and quadratic terms of time (X_1) and temperature (X_2). The reparametrized model was able to explain 94% of the total variation between the observed and predicted values. The regression was statistically significant

($F_{calculated}$ (22.1) > $F_{tabulated}$ (4.3)), whereas the F-test for lack of fit was not statistically significant ($F_{calculated}$ (2.9) < $F_{tablulated}$ (19.3)) at $p < 0.05$ (Table S2—Supplementary Material). According to the predictive model and response surface (Figure 3B), 68% free oil recovery can be achieved by chemical demulsification when using 55 °C and 46 min.

Higher temperatures can promote almond protein denaturation, leading to reduced protein solubility [47], which could help destabilize the cream emulsion. The denaturation temperature of almond proteins (globulins, albumin, and prolamins) ranges from 54 to 62 °C [20], which corresponds to the maximum demulsification yields observed in our study. Moreover, at the pH corresponding to the protein isoelectric point, electrostatic repulsion between oil droplets decreases, thus further enhancing oil droplets' coalescence and higher free oil yield [10].

3.3. Experimental Validation of Optimum Enzymatic and Chemical Demulsification Conditions

In order to validate the optimum demulsification conditions predicted by each empirical model, optimum demulsification conditions were performed in triplicate for each demulsification strategy, and a t-test was used to evaluate if the data obtained were statistically different from the predicted value.

For the enzymatic demulsification, the experimental validation was performed at pH 8.0 and 0.8% of enzyme (optimum demulsification conditions identified by the model). Enzymatic demulsification yields of 65.77 ± 2.77% were achieved during the experimental validation, not being statistically different ($p < 0.05$) from the value predicted by the model (63.11%) (Table 6). These results demonstrate that the predictive model is reliable and accurate for the prediction of how pH and amount of enzyme affect the demulsification efficiency of the AEP cream. For the chemical approach, the validation was performed at optimum conditions suggested by the regression model (55 °C and 42 min), and demulsification yields of 65.53 ± 2.29% were achieved. These values were not statistically different ($p < 0.05$) from the predicted value (68.04%) (Table 6).

Table 6. Experimental validation of optimum enzymatic and chemical demulsification conditions.

Treatment	Experimental Conditions	Demulsification Yield (%)	
		Predicted *	Experimental Validation **
Chemical	pH 5.0, 55 °C, 46 min, no enzyme	68.14 [a]	65.53 ± 2.29 [a,A]
Enzymatic	pH 8.0, 50 °C, 30 min, 0.8% (*w/w*) of enzyme	63.11 [a]	65.77 ± 2.77 [a,A]
Control	pH 9.0, 50 °C, 60 min, no enzyme	-	7.85 ± 0.51 [B]

* Predicted value by the regression model. ** Result presented as the mean ($n = 3$) ± SD, different lower letters in the same row indicate a significant difference between the predicted vs. observed value and different capital letters within the column indicate a significant difference among the demulsification treatments ($p < 0.05$).

Despite the similar demulsification efficiency of chemical and enzymatic strategies (~65%), enzymatic demulsification promoted a 29% reduction in the demulsification time (from 42 into 30 min) and a reduction in the reaction temperature used (from 55 to 50 °C), which could mitigate enzyme and energy costs. Furthermore, both strategies significantly increased the demulsification yield compared with non-optimized conditions (control). The low demulsification yield achieved by the control (7.8%) (Table 6) reflects the formation of a stable cream emulsion during the AEP, which has been attributed to the presence of proteins that act as an excellent emulsifier. Our results are in agreement with those reported by Chabrand and Glatz [35], Chabrand et al. [33], Tabtabaei and Diosady [37], and Wu et al. [10], who showed that the use of enzymatic reactions leads to greater oil droplet coalescence, thus assisting with free oil release from creams produced from soybean flour, dehulled yellow mustard, and extruded soybean flakes.

Our results are also in agreement with a study by Souza et al. [29], who reported the free oil recovery from the cream produced from the aqueous extraction of almond cake. For the enzymatic strategy, the authors reported similar demulsification yields (63% in theirs vs. 65% in ours) using less enzyme during the cream demulsification (0.5% in their study vs. 0.8% in ours). However, the incubation time was three times higher than the one used in our study (90 min vs. 30 min) at the same temperature (50 °C) and slightly different pH (9.0 in theirs vs. 8.0 in ours). For the chemical approach, our results were significantly higher (65 vs. 37%), with a significant reduction in the incubation time (46 vs. 90 min) compared with those reported by the same authors using the same temperature. The observed difference could be attributed to compositional differences in both starting materials (almond cake in theirs vs. almond flour in ours) as well as to the different temperatures to which these materials were subjected to during milling and mechanical pressing. The reduced fat content in the almond cake (16 vs. 43% for the almond flour) and the reduced oil yield in the cream (10 vs. 58% for the almond flour) indicate a potential effect of matrix composition and upstream unit operations to which the material might have been subjected to on the efficiency of specific demulsification strategies. Therefore, our results highlight the importance of careful processing optimization to maximize the overall recovery of the extracted oil.

3.4. Effects of Demulsification Strategies on the Physicochemical Properties of the Cream Proteins

3.4.1. Low Molecular Weight (MW) Polypeptide Profile Characterization of the Cream Proteins by Sodium Dodecyl Sulfate-Polyacrylamide Gel Electrophoresis (SDS-PAGE)

Demulsification yields are closely related to the structure of the protein remaining in the cream. The amount and type of proteins present at the interface significantly affect the emulsion stability [20,35]. Therefore, understanding the impact of different demulsification strategies in the cream protein structure becomes necessary to better select the approach leading to higher demulsification yields.

The effects of chemical and enzymatic demulsification strategies (CD and ED) on the polypeptide profile of the cream are shown in Figure 4A. The AEP cream contains high molecular weight amandin proteins (α- and β-chain) and smaller oleosin proteins that form a multilayer interface (intact cream, Lane 2). A similar electrophoretic profile was observed for the cream proteins after chemical demulsification (Lane 3), which was expected because the chemical strategy consists of destabilizing the cream emulsion by adjusting the pH to the protein isoelectric point. Lanes 2 and 3 show that the proteins were composed of two major bands at 38–40 kDa (~22% of the lane area) and 20 kDa (~30%). The two fragments were reported as the subunits of the amandin (α- and β-subunit), which represents 70% of the almond protein [39] (Figure 4B). However, the use of enzymes to destabilize the cream resulted in complete degradation of the basic subunit of amandin (α-subunit) and a significant reduction in the acidic subunit (β-subunit) (from 30 to 12% of relative abundance) (Lane 4) (Figure 4A,B). Approximately 70% of the peptides had molecular weight smaller than 10 kDa, demonstrating that enzymatic demulsification of the cream effectively reduced the size of the interfacial proteins. The presence of short polypeptides (<10 kDa) and the reduced amount of peptides within ~15–26 kDa after enzymatic demulsification might indicate the hydrolysis of amphiphilic oleosins small proteins that prevent the oil droplets from coalescing [15,48,49]. Oleosin hydrolysis could affect the interfacial film integrity and promote the coalescence of the oil, thus explaining the reduced stability and faster oil release of the enzymatic strategy in relation to the chemical and control treatments.

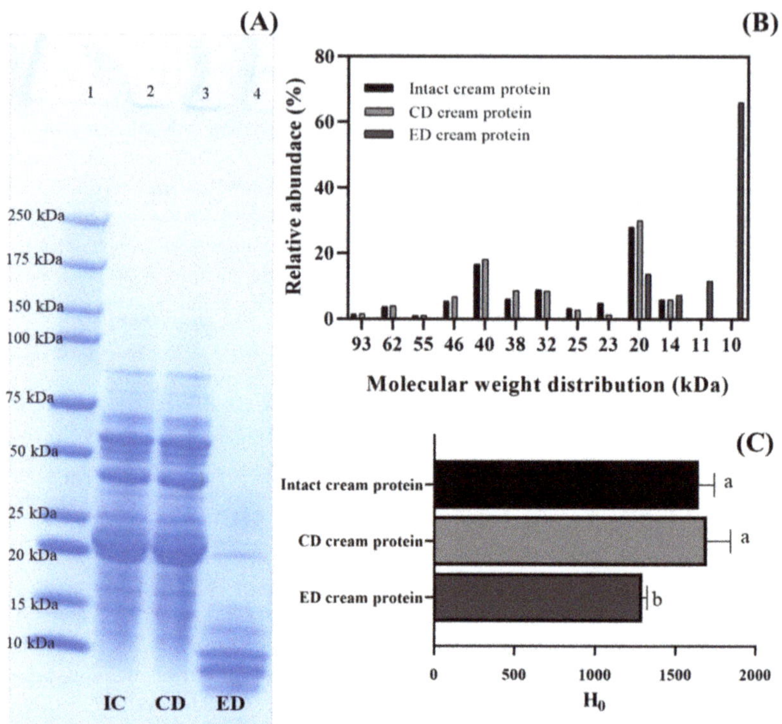

Figure 4. sodium dodecyl sulfate-polyacrylamide gel electrophoresis (SDS-PAGE) of cream proteins (**A**), Lane 1: molecular weight standard (10–250 kDa), Lane 2: intact cream (IC) protein, Lane 3: chemically demulsified (CD) cream protein, and Lane 4: enzymatically demulsified (ED) cream protein; molecular weight distribution and relative abundance of the cream proteins (**B**); protein surface hydrophobicity (H_0) of the cream proteins (**C**). Different letters indicate a significant difference ($p < 0.05$) among the samples.

3.4.2. Surface Hydrophobicity (H_0) of Cream Proteins

The physicochemical properties of proteins play a key role in determining their emulsification properties [50]. Proteins from the intact cream exhibited similar surface hydrophobicity (H_0) to those after chemical demulsification (1644.1 ± 79.9 vs. 1693.7 ± 122.3) (Figure 4C). However, the cream proteins after enzymatic demulsification had significantly lower surface hydrophobicity (1293.7 ± 25.3). A reduction in surface hydrophobicity after enzymatic hydrolysis has been reported for soybeans [51,52], peanuts [31,46], and lentil proteins [53], in agreement with the present study. Jung et al. [52] proposed that protein hydrolysis could promote the exposure of buried hydrophobic groups and their aggregation via hydrophobic interactions, consequently reburying them in larger aggregates, which could lead to reduced surface hydrophobicity. However, the exact mechanism of enzymatic hydrolysis on H_0 remains unclear. Reduced hydrophobicity was associated with reduced emulsion stability of peanut cream protein [31]. Therefore, the reduced H_0 value observed in our study after enzymatic demulsification of the cream might explain the faster destabilization of the almond cream.

3.5. Fatty Acid Composition of Almond Oil from Chemical and Enzymatic Demulsification Strategies and Solvent Extraction

The fatty acid composition of the enzymatically and chemically recovered oils and of the solvent extracted oil is reported in Table 7. Regardless of the extraction/demulsification method employed, the major fatty acids found in the almond oil were oleic (C18:1-cis, 72–74%), linoleic (C18:2, 20–22%), and palmitic acid (C16:0, 4.5–5.5%). No significant differences were observed for the fatty acid composition of the oils recovered from enzymatic and chemical demulsification and solvent extraction (Soxhlet). Stearic acid (C18:0), palmitoleic (C16:1), alpha-linolenic acid (C18:3 (*n*-3), and eicosanoic acid (C20:0) were detected in the lowest amounts, with no significant difference among the strategies used to recover/extract the oil. The oil composition is in accordance with that reported by Sathe et al. [54]. Our results are also in agreement with those reported by Latif and Anwar [55], where no significant difference was observed among sesame oils extracted with solvent (Soxhlet) and those obtained by the enzyme assisted aqueous extraction process. To the best of our knowledge, this is the first study to compare the fatty acid composition of almond oil obtained from different demulsification strategies.

Table 7. Fatty acid composition (%) of almond oil recovered from chemical and enzymatic demulsification strategies.

Fatty Acid (%, *w/w* *)	Enzymatic Demulsification	Chemical Demulsification	Solvent Extraction
C16:0	5.01 ± 0.4	5.50 ± 0.54	4.58 ± 0.36
C16:1	0.24 ± 0.02	0.28 ± 0.05	0.20 ± 0.04
C18:0	0.42 ± 0.07	0.46 ± 0.06	0.45 ± 0.04
C18:1 (cis)	71.73 ± 1.03	72.85 ± 1.38	74.37 ± 2.12
C18:2 (*n*-6)	22.55 ± 0.60	20.86 ± 1.80	20.33 ± 1.75
C18:3 (*n*-3)	0.03 ± 0.001	0.02 ± 0.001	0.05 ± 0.002
C20:0	0.03 ± 0.002	0.03 ± 0.001	0.02 ± 0.001

* Relative percentage of the total fatty acid groups. The absence of letters indicates no statistical difference ($p < 0.05$) in the row for each fatty acid analyzed.

4. Conclusions

Chemical and enzymatic demulsification strategies significantly increased the recovery of almond oil extracted by the aqueous extraction process. Although similar oil recovery (65%) was achieved by both approaches, enzymatic demulsification was accomplished in a shorter reaction time at lower temperatures, which could reduce processing costs and preserve oil quality. Enzymatic demulsification resulted in significant changes in the physicochemical properties of the cream protein. Reduced cream stability after enzymatic demulsification could be attributed to the hydrolysis of the alpha unit of amandin and reduced protein hydrophobicity. Fatty acid composition of the oils recovered by both demulsification strategies was similar to that of the hexane extracted oil. These results provide destabilization strategies for the oil-rich emulsion formed during the aqueous extraction processing of almond flour, thus contributing to the development of this environmentally friendly technology. Subsequent evaluation of the use of enzyme during the extraction is warranted to further improve the overall recovery of the extracted oil. Moreover, considering the possible effects of the enzyme on hydrolysis-induced protein unfolding provides the basis for expanding this processing approach to extract proteins from other matrices.

Supplementary Materials: The following are available online at http://www.mdpi.com/2227-9717/8/10/1228/s1. Table S1. Analysis of variance (ANOVA) of the estimated regression models for enzymatic cream demulsification for the AEP cream; Table S2. Analysis of variance (ANOVA) of the estimated regression models for chemical cream demulsification for the AEP cream.

Author Contributions: Conceptualization, J.M.L.N.d.M.B.; methodology, F.F.G.D., N.M.d.A., J.M.L.N.d.M.B., and A.Y.T.; software, F.F.G.D.; formal analysis, F.F.G.D. and T.S.P.d.S.; investigation F.F.G.D. and T.S.P.d.S.; resources, J.M.L.N.d.M.B.; writing—original draft preparation, F.F.G.D. and J.M.L.N.d.M.B.; writing—review and editing, F.F.G.D., A.Y.T., and J.M.L.N.d.M.B.; visualization, supervision, J.M.L.N.d.M.B.; project administration, J.M.L.N.d.M.B.; funding acquisition, J.M.L.N.d.M.B. All authors have read and agreed to the published version of the manuscript.

Funding: Funding for this project was made possible by the U.S. Department of Agriculture's (USDA) Agricultural Marketing Service through grant AM170100XXXXG011. Its contents are solely the responsibility of the authors and do not necessarily represent the official views of the USDA.

Conflicts of Interest: The authors declare no conflict of interest.

References

1. Almond Board of California. *Economic Impact of Almond Production*; Almond Board of California: Modesto, CA, USA, 2018.
2. Mandalari, G.; Bisignano, G.; Wickham, M.S.J.; Narbad, A. Potential Prebiotic Properties of Almond (*Amygdalus communis* L.). *Seeds* **2008**, *74*, 4264–4270. [CrossRef] [PubMed]
3. Liu, Z.; Lin, X.; Huang, G.; Zhang, W.; Rao, P.; Ni, L. Prebiotic effects of almonds and almond skins on intestinal microbiota in healthy adult humans. *Anaerobe* **2014**, *26*, 1–6. [CrossRef] [PubMed]
4. Fasoli, E.; D'Amato, A.; Kravchuk, A.; Citterio, A.; Righetti, P.G. In-depth proteomic analysis of non-alcoholic beverages with peptide ligand libraries. I: Almond milk and orgeat syrup. *J. Proteom.* **2011**, *74*, 1080–1090. [CrossRef] [PubMed]
5. Sathe, S.K. Solubilization, electrophoretic characterization and in vitro digestibility of almond (*Prunus amygdalus*) proteins. *J. Food Biochem.* **1993**, *16*, 249–264. [CrossRef]
6. Richardson, D.P.; Astrup, A.; Cocaul, A.; Ellis, P. The nutritional and health benefits of almonds: A healthy food choice. *Food Sci. Tech. Bull. Funct. Foods* **2009**, *6*, 41–50. [CrossRef]
7. Ahmad, Z. The uses and properties of almond oil. *Complement. Ther. Clin. Pract.* **2010**, *16*, 10–12. [CrossRef]
8. Savoire, R.; Lanoisellé, J.-L.; Vorobiev, E. Mechanical Continuous Oil Expression from Oilseeds: A Review. *Food Bioprocess Technol.* **2013**, *6*, 1–16. [CrossRef]
9. Johnson, L.A. Recovery of Fats and Oils from Plant and Animal Sources. In *Introduction to Fats and Oils*; Wan, P.J., Farr, W., Eds.; AOCS Press: Champaign, IL, USA, 2000; pp. 108–135.
10. Wu, J.; Johnson, L.A.; Jung, S. Demulsification of oil-rich emulsion from enzyme-assisted aqueous extraction of extruded soybean flakes. *Bioresour. Technol.* **2009**, *100*, 527–533. [CrossRef]
11. Environmental Protection Agency. *40 CFR Part 63 National Emissions Standards for Hazardous Air Pollutants: Solvent Extraction for Vegetable Oil Production, Final Rule, Federal Register*; Environmental Protection Agency: Washington, DC, USA, 2001; pp. 19005–19026.
12. de Moura, J.M.L.N.; Maurer, D.; Jung, S.; Johnson, L.A. Integrated Countercurrent Two-Stage Extraction and Cream Demulsification in Enzyme-Assisted Aqueous Extraction of Soybeans. *J. Am. Oil Chem. Soc.* **2011**, *88*, 1045–1051. [CrossRef]
13. Jung, S.; Moura, J.M.L.N.; Campbell, K.C.; Johnson, L.A. Enzyme-assisted aqueous extraction of oilseeds. In *Enhancing Extraction Processes in the Food Industry*; Lebovka, N., Vorobiev, E., Chemat, F., Eds.; CRC Press: Boca Raton, FL, USA, 2011; pp. 477–518. [CrossRef]
14. Ndlela, S.C.; Moura, J.M.L.N.; Olson, N.K.; Johnson, L.A. Aqueous Extraction of Oil and Protein from Soybeans with Subcritical Water. *J. Am. Oil Chem. Soc.* **2012**, *89*, 1145–1153. [CrossRef]
15. Rosenthal, A.; Pyle, D.L.; Niranjan, K. Aqueous and enzymatic processes for edible oil extraction. *J. Am. Oil Chem. Soc.* **1996**, *19*, 402–420. [CrossRef]
16. Campbell, K.A.; Glatz, C.E.; Johnson, L.A.; Jung, S.; Moura, J.M.N.; Kapchie, V.; Murphy, P. Advances in Aqueous Extraction Processing of Soybeans. *J. Am. Oil Chem. Soc.* **2011**, *88*, 449–465. [CrossRef]
17. Balvardi, M.; Rezaei, K.; Mendiola, J.A.; Ibáñez, E. Optimization of the Aqueous Enzymatic Extraction of Oil from Iranian Wild Almond. *J. Am. Oil Chem. Soc.* **2015**, *92*, 985–992. [CrossRef]
18. Yusoff, M.M.; Gordon, M.H.; Niranjan, K. Aqueous enzyme assisted oil extraction from oilseeds and emulsion de-emulsifying methods: A review. *Trends Food Sci. Technol.* **2015**, *41*, 60–82. [CrossRef]
19. Almeida, N.M.; Moura Bell, J.M.L.N.; Johnson, L.A. Properties of Soy Protein Produced by Countercurrent, Two-Stage, Enzyme-Assisted Aqueous Extraction. *J. Am. Oil Chem. Soc.* **2014**, *91*, 1077–1085. [CrossRef]
20. Li, P.; Zhang, W.; Han, X.; Liu, J.; Liu, Y.; Gasmalla, M.A.A.; Yang, R. Demulsification of oil-rich emulsion and characterization of protein hydrolysates from peanut cream emulsion of aqueous extraction processing. *J. Food Eng.* **2017**, *204*, 64–72. [CrossRef]
21. Jiang, L.; Hua, D.; Wang, Z.; Xu, S. Aqueous enzymatic extraction of peanut oil and protein hydrolysates. *Food Bioprod. Process.* **2010**, *88*, 233–238. [CrossRef]
22. Dev, D.K.; Quensel, E. Preparation and Functional Properties of Linseed Protein Products Containing Differing Levels of Mucilage. *J. Food Sci.* **1988**, *53*, 1834–1837. [CrossRef]

23. Tirgar, M.; Silcock, P.; Carne, A.; Birch, E.J. Effect of extraction method on functional properties of flaxseed protein concentrates. *Food Chem.* **2017**, *215*, 417–424. [CrossRef]

24. Aliakbarian, B.; De Faveri, D.; Converti, A.; Perego, P. Optimisation of olive oil extraction by means of enzyme processing aids using response surface methodology. *Biochem. Eng. J.* **2008**, *42*, 34–40. [CrossRef]

25. Najafian, L.; Ghodsvali, A.; Haddad Khodaparast, M.H.; Diosady, L.L. Aqueous extraction of virgin olive oil using industrial enzymes. *Food Res. Int.* **2009**, *42*, 171–175. [CrossRef]

26. Moreau, R.A.; Dickey, L.C.; Johnston, D.B.; Hicks, K.B. A Process for the Aqueous Enzymatic Extraction of Corn Oil from Dry Milled Corn Germ and Enzymatic Wet Milled Corn Germ (E-Germ). *J. Am. Oil Chem. Soc.* **2009**, *86*, 469–474. [CrossRef]

27. Moreau, R.A.; Johnston, D.B.; Powell, M.J.; Hicks, K.B. A comparison of commercial enzymes for the aqueous enzymatic extraction of corn oil from corn germ. *J. Am. Oil Chem. Soc.* **2004**, *81*, 1071–1075. [CrossRef]

28. Almeida, N.M.; Dias, F.F.G.; Rodrigues, M.I.; de Moura Bell, J.M.L.N. Effects of Processing Conditions on the Simultaneous Extraction and Distribution of Oil and Protein from Almond Flour. *Processes* **2019**, *7*, 844. [CrossRef]

29. Souza, T.S.P.; Dias, F.F.G.; Koblitz, M.G.B.; de Moura Bell, J.M.L.N. Effects of enzymatic extraction of oil and protein from almond cake on the physicochemical and functional properties of protein extracts. *Food Bioprod. Process.* **2020**, *122*, 280–290. [CrossRef]

30. Souza, T.S.P.; Dias, F.F.G.; Koblitz, M.G.B.; de Moura Bell, J.M.L.N. Aqueous and Enzymatic Extraction of Oil and Protein from Almond Cake: A Comparative Study. *Processes* **2019**, *7*, 472. [CrossRef]

31. Li, P.; Gasmalla, M.A.A.; Liu, J.; Zhang, W.; Yang, R.; Aboagarib, E.A.A. Characterization and demusification of cream emulsion from aqueous extraction of peanut. *J. Food Eng.* **2016**, *185*, 62–71. [CrossRef]

32. Moura, J.M.L.N.; Almeida, N.M.; Johnson, L.A. Scale-up of Enzyme-Assisted Aqueous Extraction Processing of Soybeans. *J. Am. Oil Chem. Soc.* **2009**, *86*, 809–815. [CrossRef]

33. Chabrand, R.M.; Kim, H.J.; Zhang, C.; Glatz, C.E.; Jung, S. Destabilization of the emulsion formed during aqueous extraction of Soybean oil. *J. Am. Oil Chem. Soc.* **2008**, *85*, 383–390. [CrossRef]

34. Damodaran, S.; Anand, K. Sulfhydryl–Disulfide Interchange-Induced Interparticle Protein Polymerization in Whey Protein-Stabilized Emulsions and Its Relation to Emulsion Stability. *J. Agric. Food Chem.* **1997**, *45*, 3813–3820. [CrossRef]

35. Chabrand, R.M.; Glatz, C.E. Destabilization of the emulsion formed during the enzyme-assisted aqueous extraction of oil from soybean flour. *Enzym. Microb. Technol.* **2009**, *45*, 28–35. [CrossRef]

36. Moura, J.M.L.N.; Campbell, K.; Mahfuz, A.; Jung, S.; Glatz, C.E.; Johnson, L. Enzyme-Assisted Aqueous Extraction of Oil and Protein from Soybeans and Cream De-emulsification. *J. Am. Oil Chem. Soc.* **2008**, *85*, 985–995. [CrossRef]

37. Tabtabaei, S.; Diosady, L.L. Aqueous and enzymatic extraction processes for the production of food-grade proteins and industrial oil from dehulled yellow mustard flour. *Food Res. Int.* **2013**, *52*, 547–556. [CrossRef]

38. de Aquino, D.S.; Fanhani, A.; Stevanato, N.; da Silva, C. Sunflower oil from enzymatic aqueous extraction process: Maximization of free oil yield and oil characterization. *J. Food Process Eng.* **2019**, *42*, e13169. [CrossRef]

39. Sathe, S.K.; Wolf, W.J.; Roux, K.H.; Teuber, S.S.; Venkatachalam, M.; Sze-Tao, K.W.C. Biochemical Characterization of Amandin, the Major Storage Protein in Almond (*Prunus dulcis* L.). *J. Agric. Food Chem.* **2002**, *50*, 4333–4341. [CrossRef]

40. Laemmli, U.K. Cleavage of structural proteins during the assembly of the head of bacteriophage T4. *Nature* **1970**, *227*, 680–685. [CrossRef]

41. Zhang, Y.; Zhao, W.; Yang, R.; Ahmed, M.A.; Hua, X.; Zhang, W.; Zhang, Y. Preparation and functional properties of protein from heat-denatured soybean meal assisted by steam flash-explosion with dilute acid soaking. *J. Food Process Eng.* **2013**, *119*, 56–64. [CrossRef]

42. American Oil Chemists' Society AOCS Standard Procedure Am 5-04. Rapid Determination of Oil/Fat Utilizing High-Temperature Solvent Extraction. In *Official Methods and Recommended Practices of AOCS*; American Oil Chemists' Society: Urbana, IL, USA, 2017.

43. Association of Official Analytical Chemists. Official Method 922.06 Fat in Flour. Acid Hydrolysis Method. In *Official Methods of Analysis of AOAC International*; Association of Official Analytical Chemists: Rockville, MD, USA, 2012.

44. AACC Approved Methods of Analysis, 11th Edition—AACC Method 44-40.01, Moisture—Modified Vacuum-Oven Method. Available online: http://methods.aaccnet.org/summaries/44-40-01.aspx (accessed on 6 August 2020).

45. Jung, S. Aqueous extraction of oil and protein from soybean and lupin: A comparative study. *Food Process. Preserv.* **2009**, *33*, 547–559. [CrossRef]

46. Zhang, S.B.; Lu, Q.Y. Characterizing the structural and surface properties of proteins isolated before and after enzymatic demulsification of the aqueous extract emulsion of peanut seeds. *Food Hydrocoll.* **2015**, *47*, 51–60. [CrossRef]

47. Kinsella, J.E. Functional properties of soy proteins. *J. Am. Oil Chem. Soc.* **1979**, *56*, 242–258. [CrossRef]

48. Huang, A.H.C. Oil bodies and oleosins in seeds. *Annu. Rev. Plant Physiol. Plant Mol. Biol.* **1992**, *43*, 177–200. [CrossRef]

49. Beisson, F.; Ferté, N.; Voultoury, R.; Arondel, V. Large scale purification of an almond oleosin using an organic solvent procedure. *Plant Physiol. Biochem.* **2001**, *39*, 623–630. [CrossRef]

50. Ghribi, A.M.; Gafsi, I.M.; Sila, A.; Blecker, C.; Danthine, S.; Attia, H.; Bougatef, A.; Besbes, S. Effects of enzymatic hydrolysis on conformational and functional properties of chickpea protein isolate. *Food Chem.* **2015**, *187*, 322–330. [CrossRef] [PubMed]

51. Surówka, K.; Zmudziński, D.; Surówka, J. Enzymatic modification of extruded soy protein concentrates as a method of obtaining new functional food components. *Trends Food Sci. Technol.* **2004**, *15*, 153–160. [CrossRef]

52. Jung, S.; Murphy, P.A.; Johnson, L.A. Physicochemical and Functional Properties of Soy Protein Substrates Modified by Low Levels of Protease Hydrolysis. *J. Food Sci.* **2005**, *70*, C180–C187. [CrossRef]

53. Avramenko, N.A.; Low, N.H.; Nickerson, M.T. The effects of limited enzymatic hydrolysis on the physicochemical and emulsifying properties of a lentil protein isolate. *Food Res. Int.* **2013**, *51*, 162–169. [CrossRef]

54. Sathe, S.K.; Seeram, N.P.; Kshirsagar, H.H.; Heber, D.; Lapsley, K.A. Fatty acid composition of California grown almonds. *J. Food Sci.* **2008**, *73*. [CrossRef]

55. Latif, S.; Anwar, F. Aqueous enzymatic sesame oil and protein extraction. *Food Chem.* **2011**, *125*, 679–684. [CrossRef]

Article

Optimization and Validation of Rancimat Operational Parameters to Determine Walnut Oil Oxidative Stability

Lucía Félix-Palomares and Irwin R. Donis-González *

Department of Biological & Agricultural Engineering, University of California, Davis, One Shields Ave., Davis, CA 95616, USA; lfelix@ucdavis.edu
* Correspondence: irdonisgon@ucdavis.edu; Tel.: +1-530-752-8986

Abstract: This study was performed to optimize and validate Rancimat (Metrohm Ltd., Herisau, Switzerland) operational parameters including temperature, air-flow, and sample weight to minimize Induction-Time (*IT*) and *IT*-Coefficient-of-Variation (*CV*), using Response Surface Methodology (RSM). According to a Box–Behnken experimental design, walnut oil equivalent to 3-, 6-, or 9-g was added to each reaction vessel and heated to 100, 110, or 120 °C, while an air-flow equal to 10-, 15-, or 20-L·h^{-1} was forced through the reaction vessels. A stationary point was found per response variable (*IT* and *CV*), and optimal parameters were defined considering the determined stationary points for both response variables at 100 °C, 25 L·h^{-1}, and 3.9 g. Optimal parameters provided an *IT* of 5.42 ± 0.02 h with a *CV* of 1.25 ± 0.83%. RSM proved to be a useful methodology to find Rancimat operational parameters that translate to accurate and efficient values of walnut oil *IT*.

Keywords: walnut oil; rancidity; induction time; rancimat; response surface methodology

Citation: Félix-Palomares, L.; Donis-González, I.R. Optimization and Validation of Rancimat Operational Parameters to Determine Walnut Oil Oxidative Stability. *Processes* **2021**, *9*, 651. https://doi.org/10.3390/pr9040651

Academic Editor: Blanca Hernández-Ledesma

Received: 25 March 2021
Accepted: 6 April 2021
Published: 8 April 2021

Publisher's Note: MDPI stays neutral with regard to jurisdictional claims in published maps and institutional affiliations.

1. Introduction

Oxidative deterioration, commonly known as lipid oxidation or rancidity, causes off-flavors, discoloration, unpleasant odors, and decreases edible oil nutritional value. Lipid oxidation is a problem of economic concern to the walnut industry, as it decreases the nuts positive quality characteristics (e.g., flavor, smell, and color) and shortens their shelf life [1]. Walnuts are particularly susceptible to oxidation-induced rancidity because of their high oil content (60 to 70%) and elevated content (~70%) of polyunsaturated fatty acids (PUFA), such as linoleic (56.5–57.5%) and linolenic (11.6–13.2%) fatty acids, and monounsaturated (~21%) fatty acids, mainly composed by including oleic acid (21.0–21.2%) [2,3]. Nuts with lower PUFA content have a longer shelf life [4] like hazelnut and macadamia, which contain around 13.6% and 3.7%, respectively [5].

In order to evaluate rancidity in walnut oil, extracted from fresh walnut kernels, it is necessary to determine the oil oxidative stability. Walnut rancidity in fresh walnut samples is highly variable [6]. Therefore, it is preferred to estimate the oil rancidity, instead of its corresponding fresh ground samples, as oil is a better and more uniform representation of a large batch of walnuts. The quantification of free fatty acids (FFA) and peroxides or peroxide value (PV), p-Anisidine, and test of thiobarbituric acid (TBA) are important measurements that have been used for decades to evaluate oil oxidative stability [1]. However, these tests provide information regarding a temporal fat oxidative state, and rely on the use of toxic solvents [7] and manual titration that may lead to erroneous results [8,9]. Considering that lipid oxidation is a dynamic process, a forward path in directly evaluating rancidity is to use an accelerated oxidation method to determine the lipid's relative oxidative stability, also known as the Induction Time (*IT*), which is defined as the resistance of lipids to be oxidized [9]. The Rancimat equipment is widely used by the edible and pharmaceutical oils industries, as it does not require the use of hazardous materials and intensive labor in comparison to other accelerated oxidation methods, such as the Active Oxygen Method (AOCS Method Cd 12-57) [9,10].

The Rancimat equipment (Metrohm Ltd., Herisau, Switzerland) automatically determines the *IT* by measuring conductivity changes as volatile compounds are generated through an accelerated oxidation method or aging test. The Rancimat accelerated oxidation method consists of exposing an oil sample to a constant high temperature (between 50 and 200 °C), and airflow (between 1 and 25 L·h^{-1}) to guarantee a sufficient oxygen supply to rapidly induce lipid oxidation. Operational parameters are typically established by the Rancimat equipment manufacturer, suggesting that parameters to estimate walnut oil *IT* are 3 g of sample, using a temperature of 110 °C, and an airflow equal to 20 L·h^{-1} [11]. Nonetheless, others studies that have evaluated walnut oil *IT*, using the Rancimat equipment, have used a broad range of sample weights (2.5 to 5 g), temperatures (100 to 110 °C), and airflows (15 to 20 L·h^{-1}) [2,4]. Unfortunately, none of the previous studies including the suggested manufacturer parameters properly describe how the operational parameters were optimized and validated. Therefore, results are widely variable, inconsistent, and incomparable. Therefore, there is a clear need to accurately describe how walnut Rancimat equipment *IT* determination can be optimized and validated, as described in this study. Additionally, there are currently no studies properly describing the effect of differences in the Rancimat operational parameters in the *IT* determination. This study provides an in-depth understanding of this effect, providing valuable information to those in need of properly estimating walnut oil oxidative stability.

Response Surface Methodology (RSM) is a multivariate statistic technique used to develop and improve processes in which several input variables may have an effect in the process performance and output [12]. RSM statistical designs offer quadratic response surfaces, which include central composite, Doehlert matrix, three level full factorial, and Box–Behnken designs. Overall, the Box–Behnken design has proven to be one of the most efficient, as it requires a fewer number of experiments than the minimum required for a complete three level factorial design [13]. By using Box–Behnken experimental design, Donis-González, Guyer [14] successfully optimized the computed tomography image scanning parameters (voltage, current, and slice thickness) to predict chestnut (*Castanea* spp.) internal quality. Rodríguez-González, Femenia [15] also used a Box–Behnken experimental design to resolve the parameters (plant age, pasteurization temperature, and time) that maximized alcohol swelling, water retention capacity, and fat absorption capacity of *Aloe vera* as a polysaccharide-rich food ingredient extract.

Therefore, the objective of this study was to use the RSM statistical technique in combination with a Box–Behnken experimental design to optimize and validate the Rancimat operational parameters (sample weight, temperature, and airflow), and establish a standard for future walnut oil *IT* estimations. The goal was to minimize the *IT* Coefficient-of-Variation (*CV*) within the same oil sample, and report the yielded *IT* measurements. A minimized *CV* within the same oil is reflected as a minimum error in *IT* estimation.

2. Materials and Methods

Figure 1 shows a schematic representation of the methodology followed in this study.

2.1. Walnut Collection, Oil Extraction and Storage

Walnuts (*Juglans regia* L.) var. "Howards" and "Tulare" were collected during the 2017 harvesting season from two walnut hullers/driers located at Meridian and Visalia, California. Walnuts were dehydrated in a six-column convective dryer at 43 °C for 48 h, to achieve an in-shell moisture content of around 6 to 8% wet basis (MC$_{wb}$), as specified by Khir, Pan [16]. In-shell walnuts were stored in burlap bags at ambient conditions (8–20 °C, and 53–93% relative humidity) for six months, before oil extraction.

Six kg of walnuts were manually shelled using a walnut cracker, and shells were disposed of. Walnut kernels were placed inside Ziploc bags, manually crushed with a hammer, and strained through a 1.70 mm sieve. Forty g batches of the crushed walnut kernels retained on a secondary sieve (710 μm) were placed in a stainless-steel oil extraction cylinder (5.7 cm diameter by 7.6 cm height) to extract their oil. Walnut oil was extracted using a Carver Press Model 3351 by applying a constant force of 5 metric tons (MT) for 5 min (Carver Inc., Wabash, IN, USA). On average, 15 mL of walnut oil were collected per batch in a stainless-steel pan.

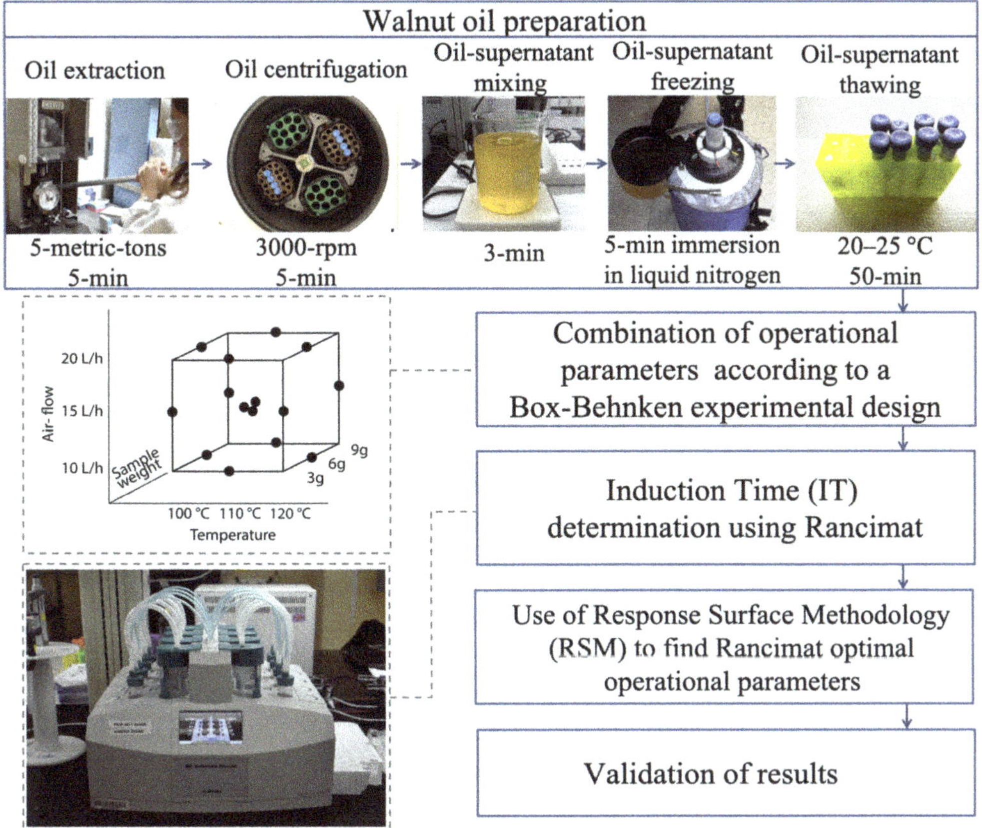

Figure 1. Description of the followed methodology to optimize Rancimat operational parameters.

For this study, oil extraction was repeated for approximately 70 batches, producing a total of 1 L of oil. The extracted oil was divided into 50 mL centrifuge tubes, and centrifuged at 3000 rpm for 5 min to precipitate the solid particles from the supernatant. Oil supernatant was collected to avoid the potential influence of solid particles on *IT* estimation. To ensure uniformity, the oil was poured in a 2 L glass beaker and stirred for 3 min on a stirring plate. Thereafter, the walnut oil was divided into 15 mL polypropylene vials and immediately immersed in liquid nitrogen for 5 min and stored at −80 °C for further analysis.

2.2. Optimization of Rancimat Operational Parameters (Independent Variables) Using Response Surface Methodology (RSM) and Box-Behnken Experimental Design

RSM using a three factor, three level Box–Behnken statistical design was used to optimize the Rancimat operational parameters (sample weight, temperature, and airflow) by maximizing the CV^{-1} [14]. In parallel, a model for the effect of operational parameters on the *IT* was also generated. The latter was performed to offer an in-depth understanding of the effect of optimized operational parameters in the walnut oil *IT*, and their implementation feasibility. R (V1.3.1093, R Development, Core Team, Vienna, Austria) software was used to develop the models. The models with coded-factors for the experimental design is described in Equation (1).

$$Y_n = b_0 + b_1X_1 + b_2X_2 + b_3X_3 + b_{12}X_1X_2 + b_{13}X_1X_3 + b_{23}X_2X_3 + b_{11}X_1{}^2 + b_{22}X_2{}^2 + b_{33}X_3{}^2 \tag{1}$$

where the index n in Y refers to each response (dependent) variable ($Y_1 = CV^{-1}$ and $Y_2 = IT$) b_0 is the regression intercept; b_1 through b_3 are the regression coefficients; and X_1, X_2, and X_3 are the Rancimat operational parameters (sample weight, temperature, and airflow) or independent variables [12].

Levels for each operational parameter were coded as low (−1), medium (0), and high (+1), as described in Myers, Montgomery [12] (Table 1a). Table 1b shows the combinations of operational parameters on each run of the Box–Behnken statistical design.

Table 1. (a) Box–Behnken statistical design factors. (b) Combination of factors in Box–Behnken statistical design.

(a)			
		Levels (coded)/uncoded	
Operational parameters (units)	Low (−1)	Medium (0)	High (+1)
Sample weight (g)	3	6	9
Temperature (°C)	100	110	120
Airflow (L·h^{-1})	10	15	20

(b)			
	Block 1		
	Independent variables (operational parameters)		
Run	Sample weight (g)	Temperature (°C)	Airflow (L·h^{-1})
1	3	110	10
2	3	100	15
3	3	120	15
4	3	110	20
5	6	100	10
6	6	120	10
7	6	110	15
8	6	110	15
9	6	110	15
10	6	100	20
11	6	120	20
12	9	110	10
13	9	100	15
14	9	120	15
15	9	110	20
	Block 2		
	Independent variables (operational parameters)		
Run	Sample weight (g)	Temperature (°C)	Airflow (L·h^{-1})
16	3	100	10
17	9	120	20

2.3. Walnut Oil IT and CV (RSM Models Dependent Variables)

Walnut oil *IT* was evaluated using a Metrohm Rancimat Model 892 (Metrohm Ltd., Herisau, Switzerland). Eight vials containing walnut oil were removed at a time from storage (−80 °C), as this is the maximum number of samples that Rancimat can analyze at the same time. Before analysis, closed vials were thawed at room temperature (between 20 to 25 °C) for approximately 50 min. According to the Box–Behnken experimental design (Table 1a,b), walnut oil equivalent to 3, 6, or 9 g was added to each reaction vessel, and heated to 100, 110, or 120 °C, while 10, 15, or 20 L·h^{-1} of filtered air was forced through the oil within the reaction vessels. Each operational parameter combination (run) was carried out in triplicate (n). Glass and plastic vessels were used and cleaned as described by the Rancimat manufacturer Metrohm [17]. The *IT CV* was calculated per run, as specified in Farhoosh [18], using Equation (2).

$$V = \frac{\sqrt{\sum_{i=1}^{n}\left(\left|IT - \frac{\sum_{i=1}^{n} IT}{n}\right|\right)^2}}{\frac{\sum_{i=1}^{n} IT}{n}} \tag{2}$$

where *CV* is the Coefficient-Of-Variation of the *IT* measurements per run (Table 1b), and n is the number of replications per run (3).

Values for the *CV* were expected to be near zero, as the oil samples and corresponding replicates were obtained from the same batch of oil and were exposed to the same handling and storage conditions. To facilitate data visualization and statistical analysis, the response variable (*CV*) was inversed (CV^{-1}).

2.4. Validation of Optimal Rancimat Operational Parameters

To validate the RSM model optimization results, an independent test was ran, in duplicate, using the optimized operational parameters (sample weight, temperature, and airflow) that correspond to the RSM stationary point as the CV^{-1} was maximized (validations-1). In addition, an independent test was run, in duplicate, after the effect of the optimized parameters onto the *IT* were considered (validations-2).

The same walnut oil was used to optimize and perform model validation-1 and -2. Ultimately, a paired *t*-test at 95% confidence was performed to evaluate if there was a significant difference in the *CV* and *IT* between validation-1 and -2. R (V1.3.1093, R Development, Core Team, Vienna, Austria) software was used to perform the *t*-test.

3. Results and Discussion

3.1. Optimization of Rancimat Operational Parameters (Independent Variables)

As shown in Table 2, the results for the lack of fit test were non-significant (p value > 0.05), suggesting that for each model (RSM for CV^{-1} and *IT*), the second order non-linear models adequately described the relationship between the operational parameters (sample weight, temperature, and airflow) and the response variables (*IT* and CV^{-1}) [12]. This is an indicator that the RSM stationary points are relevant and models were properly developed [19].

Table 2. Regression coefficients and *p*-value [a] for each non-linear response variable model.

Term (Code)	CV^{-1}		*IT* (h)	
	Coeff.	*p*-Value	Coeff.	*p*-Value
Intercept (b_0)	1.1908	**0.0110 a**	2.7166	**4.137 × 10^{-10} a**
Block effect	0.8066	0.2939	0.0255	0.7557
Sample weight (X_1)	−0.0306	0.8746	0.0165	0.4565
Temperature (X_2)	−0.2111	0.2999	−1.9661	**9.570 × 10^{-11} a**
Airflow (X_3)	0.1911	0.3442	0.0362	0.1330
Temperature x sample weight (X_2X_1)	−0.0337	0.9093	−0.0025	0.9400
Airflow x sample weight (X_3X_1)	−0.2764	0.3685	0.0117	0.7248
Temperature x airflow (X_2X_3)	−0.3699	0.2410	−0.0317	0.3572
Sample weight (X_1^2)	−0.4239	0.2020	0.0086	0.8025
Temperature (X_2^2)	−0.2150	0.4948	0.6221	**1.475 × 10^{-6} a**
Airflow (X_3^2)	−0.2759	0.3872	0.0279	0.4319

[a] ANOVA for each term at p = 0.05. Statistically significant values (p-value ≤ 0.05) are in bold text.

The regression coefficients for the CV^{-1} optimization model (Table 2) suggested that none of the operational parameters (sample weight, temperature, and airflow) or interaction of these significantly contributed to the model, as the p-values for all terms were significant (p-values ≤ 0.05). The mathematical relationship between the response (dependent variable) and operational parameters can be understood by interpreting the regression coefficients of the polynomial equation (Table 2) [14]. The coefficients represent the expected change in the response per unit change in each operational parameter (independent variable), when the remaining operational parameters are held constant [12].

Table 3 displays the results of ANOVA for the CV^{-1} model, where an R^2 of 57.8% and an R^2_{adj} of −12.4% were obtained. According to Myers, Montgomery [12], when R and R^2_{adj} differ dramatically, there is a high probability that non-significant terms have been included into the model [12]. The fact that none of the regression coefficients were significant could explain why R^2 and R^2_{adj} values differ substantially. The previous also indicates that the CV is highly variable within the range of evaluated parameters. Table 3 shows that for the CV^{-1} model none of the linear, quadratic, and polynomial order terms were significant as the three terms yielded a p-values > 0.05. This confirms that none of the operational parameters significantly contributed to the CV^{-1} model, and reiterates that the Rancimat equipment performs with high accuracy and consistency, regardless of operational parameters setup [12]. Nevertheless, a CV^{-1} maximum stationary point was discerned, as an indicator of low variation within a specific range of operation parameters, as three negative eigenvalues were obtained (Table 4) [12]. The optimized response (lowest variation) for CV^{-1} was established at 3.9 g for sample weight, 85.5 °C for temperature, and 26.7 L·h^{-1} for airflow. The stationary point for sample weight was inside the experimental region, while the stationary points for temperature and airflow were outside the experimental region. In RSM, it is possible to find a stationary point that is beyond the experimental region. This is called a rising or falling ridge. Therefore, if possible, further experiments need to be explored following the ridge path.

Table 3. ANOVA table for the inverse of coefficient of variation (CV^{-1}), induction time (IT), and non-linear response models.

Term	df	SS CV^{-1}	SS IT	MS CV^{-1}	MS IT	F-Value CV^{-1}	F-Value IT	p-Value [a] CV^{-1}	p-Value [a] IT
Block	1	0.159	0.170	0.159	0.170	0.492	41.950	0.509	**0.0064**
Linear	3	0.661	38.261	0.220	12.753	0.681	3144.644	0.594	**5.616 × 10^{-10}** [a]
Quadratic	3	0.857	0.005	0.285	0.0015	0.883	0.378	0.500	0.772
Polynomial	3	0.985	1.436	0.328	0.4787	1.015	118.029	0.449	**1.006 × 10^{-5}** [a]
Residuals	6	0.941	0.024	0.323	0.0041				
Lack of fit	4	0.865	0.018	0.216	0.0044	0.402	1.293	0.801	0.478
Pure error	2	1.075	0.007	0.537	0.0034				

[a] ANOVA for each term at p-value = 0.05. Statistically significant values (p-value ≤ 0.05) are in bold text.

Table 4. Stationary maximized points in original units for each non-linear response variable model.

Response Variable	Operational Parameters Sample Weight (g)	Operational Parameters Temperature (°C)	Operational Parameters Airflow (L·h^{-1})	Eigenvalues Sample Weight (g)	Eigenvalues Temperature (°C)	Eigenvalues Airflow (L·h^{-1})
CV^{-1}	3.9	85.5	26.7	−0.550	−0.041	−0.323
IT	2.8	125.9	17.3	0.006	0.622	0.029
Mean	3.3	105.7	22.0			
Max	3.9	125.9	26.7			
Min	2.8	85.5	17.3			
Optimized	3.9	100	25			

The surface regression plots in Figure 2a–c show that at a lower temperature and higher airflow, the CV^{-1} increases, therefore decreasing the CV. This indicates that the conditions under which walnut oil oxidizes are more variable as temperature increases and airflow decreases [18]. A rising ridge in the direction of a lower temperature and a higher airflow can also be observed, indicating further analysis must be performed on those directions [12,19]. However, the maximum airflow set point for the Rancimat equipment is 25 L·h^{-1} and it is impossible to perform studies above that value. Regardless, this study demonstrated that at the optimized stationary point for a high sample weight

(3.9 g), low temperature (85.5 °C), and high airflow (26.7 L·h^{-1}) air has the potential to evenly saturate the walnut oil, yielding a less variable *IT* [18]. This result agrees with the reported by Farhoosh [18], and Jebe and Matlock [20] that concluded that a reduction in temperature leads to a decrease in the *IT CV*.

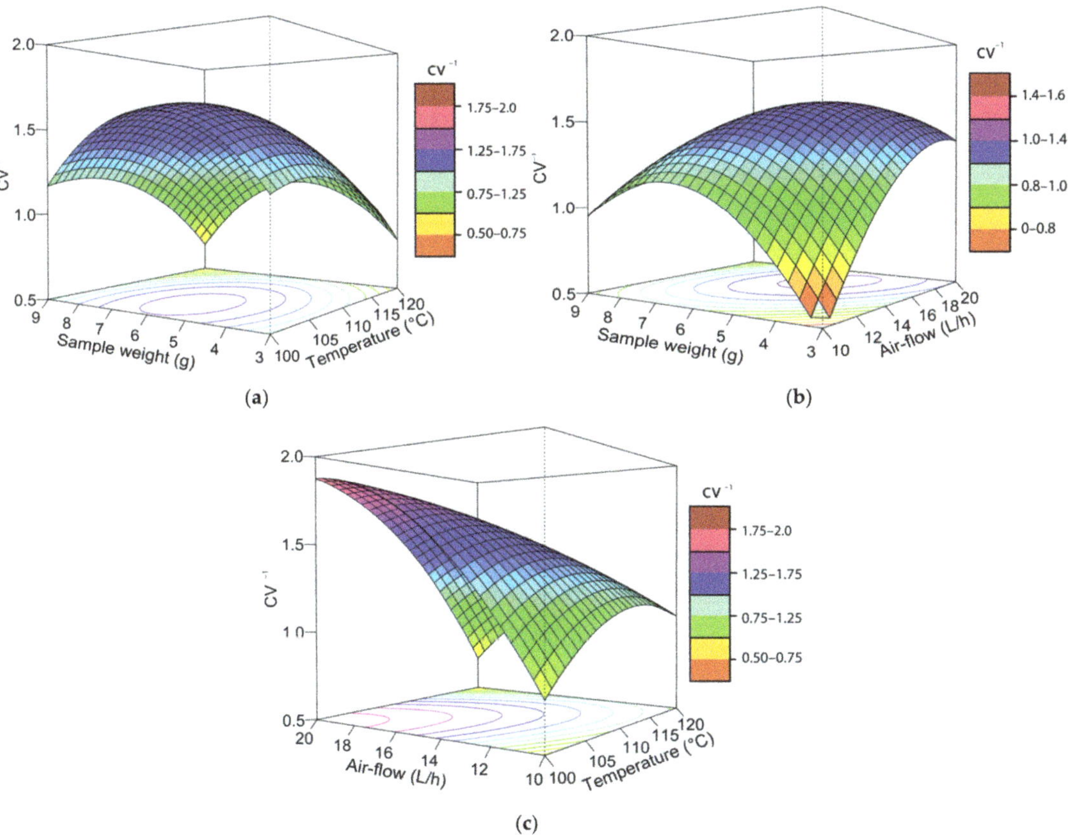

Figure 2. Surface and contour plots for the inverse of the coefficient of variation (CV^{-1}) versus (**a**) sample weight (g) and temperature (°C); (**b**) sample weight (g) and airflow (L·h^{-1}); and (**c**) airflow (L·h^{-1}) and temperature (°C).

If optimized conditions to minimize the *CV* are considered, it can be predicted that the *IT* will equal 21.31 ± 0.80 h. The previous was concluded, based on the generated *IT* model, as the *IT* at the lowest evaluated temperature (100 °C) equaled 5.33 ± 0.20 h. *IT* at the optimized conditions, which reflects the time of each test, is high and not practical for researcher and industrial application, especially as many samples need to be evaluated at once. In addition; even though a stationary point for the CV^{-1} optimized model was resolved, because none of the operational parameters (sample weight, temperature, and airflow) or interaction of these significantly contributed to the model, it is required to further explore the effect of operation parameters on the *IT* model.

Regarding the *IT* model, the *p*-value for the temperature regression coefficient was <0.05, suggesting that the effect of temperature in the *IT* was statistically significant. Sample weight, and airflow appeared to be non-significant, and no interaction of parameters was significant (*p*-value > 0.05). The temperature regression coefficient yielded a negative value, indicating that temperature has an antagonistic effect in the response, or the *IT* significantly decreases at a higher temperature. The *IT* response surface plots (Figure 3a–c) visually confirm that temperature was the only variable that significantly affected the *IT* and that the *IT* decreases with an increase in temperature. This behavior was expected, as it is known that the rate of an oxidation reaction is exponentially related to temperature [9,21]. These results coincided with those obtained by [10,18,21], who found a similar relationship between the *IT* and temperature.

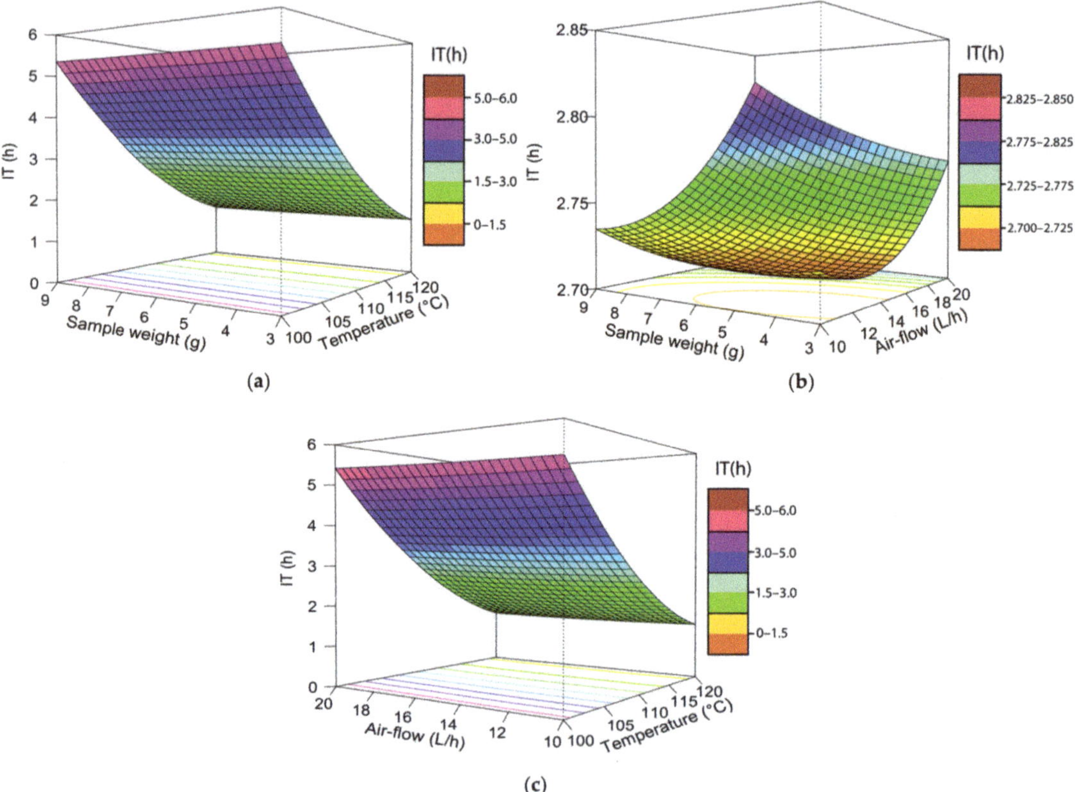

Figure 3. Surface and contour plots for the induction time (*IT*-h) versus (**a**) sample weight (g) and temperature (°C), (**b**) sample weight (g) and airflow (L·h^{-1}), and (**c**) air-flow and temperature.

The results of the ANOVA for *IT* are shown in Table 3. It was determined that the R^2 and R^2_{adj} (adjusted for degrees of freedom) were 99.9% and 99.8%, respectively. This means that the operational parameters (sample weight, temperature, and airflow) significantly explained the variation in the *IT* [14]. Moreover, the *p*-value for the first and polynomial order terms were <0.05. Therefore, the hypothesis that none of the parameters significantly contribute to the model was rejected [12]. In other words, the canonical analysis or relationship between the response and operational parameters was relevant [12,14].

For the *IT* model, a ridge is present in the sample weight equal to 2.8 g, and a temperature of 125.9 °C, advising that sample sizes below 2.8 g, and temperatures above 125.9 °C should be explored. The Rancimat upper achievable temperature limit is 120 °C, therefore, it is not possible to perform an experiment at temperatures above 125.9 °C. Moreover, previous studies have determined that temperature stabilization is unreliable at a sample below 2.5 g. It is hypothesized that air-saturation cannot be maintained for a low sample weight, when exposed to high airflow rates [10,20]. Figure 3b shows that the *IT* increases from 2.72 to 2.82 h, as the sample weight and airflow increase to their maximum values. A potential reason is that as the sample weight increases, the airflow becomes limited, making it impossible to maintain the air-saturated conditions, as required to uniformly induce the oil oxidation [10]. For these reasons, it was concluded that exploring additional conditions for the *IT* model was not feasible and will not lead to a significant improvement. Hence, if only the *IT* model was to be considered, the *IT* would be at its minimum when a sample weight, temperature, and airflow equaled 2.8 g, 125.9 °C, and 17.4 L·h^{-1}, respectively.

Studying and analyzing the relationship between operation parameters on both response variables (CV^{-1} and *IT*) was crucial to properly optimize the Rancimat method to accurately and feasible evaluate walnut oil *IT*. Optimized operational parameters need to consider the Rancimat equipment capabilities, and analysis time, while minimizing the *IT CV* (model that maximizes the CV^{-1}). Therefore, final optimized parameters for sample weight, temperature, and airflow were established at 3.9 g, 100 °C, and 25 L·h^{-1}, respectively (Table 4). Table 4, shows that the final optimized parameters are close to the arithmetic mean of operational parameters that minimize the *IT*, which ultimately defines the sample analysis time, and maximize the CV^{-1} (sample weight = 3.3 g, temperature = 105 °C, and airflow = 22 L·h^{-1}). As it was determined through the *IT* RSM model that the sample weight and airflow do not have a significant effect on the *IT*, values for these two parameters were selected whenever they minimize the *CV*, while falling within the experimental region and the Rancimat equipment capabilities.

3.2. Validation of optimal Rancimat Operational Parameters

As shown in Table 5, a *CV* of 0.966 ± 0.851%, and an *IT* of 16.261 ± 0.360 h were observed, when applying the operational parameters that exclusively maximized the CV^{-1} (validation-1). On the other hand, the final operational parameters yielded a *CV* and an *IT* equal to 1.259 ± 0.838 and 5.421 ± 0.016, respectively. The *CV* obtained in validation-1 and validation-2 were not significantly different (p-value > 0.05, df = 1.99, t-statistic = −0.34), while the *IT* was significantly different (p-value ≤ 0.05, df = 1.00, t-statistic = 42.46). These results are illustrated in Figure 4. The *IT* from validation 2, is approximately a third of the *IT* from validation-1, without significantly increasing the results variability (*CV*). A lower *IT*, as observed in the final operational parameters, represents an advantage as a significantly larger number of samples can be evaluated within the same timeframe.

Table 5. Mean and standard deviation of the *IT* (h), *CV* (%), and CV^{-1} obtained from the validation of operational parameters at the stationary point (Validation-1) and optimal parameters (Validation-2).

Validation	Sample Weight (g)	Temperature (°C)	Air Flow (L·h^{-1})	IT (h)	CV (%)	CV^{-1}
Validation-1 (Validation at the maximized stationary point for CV^{-1})	3.9	85	25	16.261 ± 0.360	0.966 ± 0.851	1.688 ± 1.486
Validation-2 (Validation at final optimal parameters)	3.9	100	25	5.421 ± 0.016	1.259 ± 0.838	1.020 ± 0.679

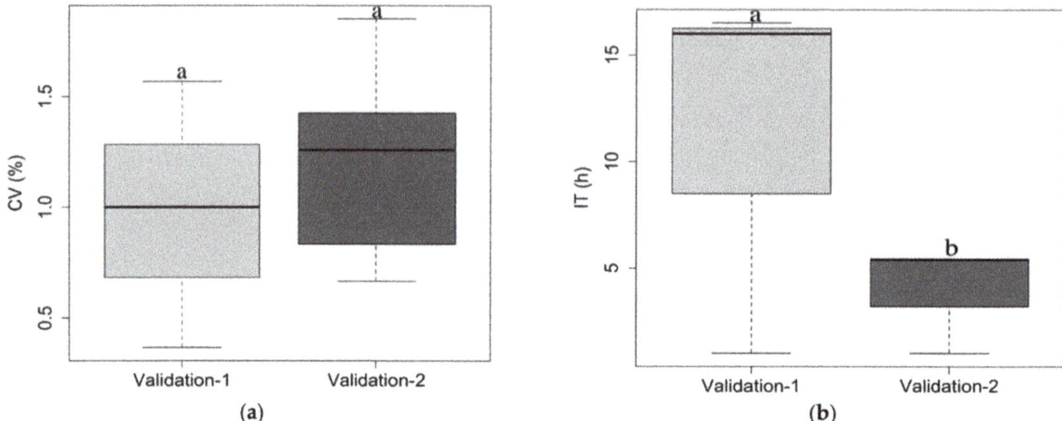

Figure 4. Box-plots showing: (**a**) The *CV* (%); and (**b**) *IT* (h) distributions in Validation-1 and Validation-2. The median is represented as a thick horizontal black line, upper and lower quartiles as a box with the maximum and minimum measurements as lines protruding from these. Box-plots followed by a different letter within each boxplot set are significantly different between each other (*p*-value $\leq$ 0.05).

4. Conclusions

Using the second-order RSM prediction models for each individual response variable (CV^{-1} and *IT*) yielded an optimized combination of Rancimat operational parameters for sample weight, temperature, and airflow equal to 3.9 g, 100 °C, and 25 L·h^{-1}, respectively. Final operational parameters yielded a *CV* of 1.259 $\pm$ 0.838%, and *IT* of 5.421 $\pm$ 0.016 h. RSM proved to be a useful methodology to properly optimize the Rancimat operational parameters (sample weight, temperature, and air flow) that allow us to precisely, accurately, and efficiently determine walnut oil *IT*. Choosing the appropriate temperature level is pivotal in determining walnut oil *IT* using the Rancimat equipment, while the airflow rate and sample size proved to not have a significant effect.

Even though the Rancimat equipment evaluates the relative oxidative stability of walnut oil, the application of the optimized operational parameters, as a standard method for future walnut oil *IT* determination, can lead to comparable results between studies. In addition, having access to a reliable and accurate method of determining walnut rancidity will facilitate the application of alternative postharvest and food processing techniques to improve walnut quality, lipid stability, and nutritional value.

Supplementary Materials: The following are available online at https://www.mdpi.com/article/10.3390/pr9040651/s1.

Author Contributions: Experimental studies, investigation, validation, and writing—original draft preparation, L.F.-P.; conceptualization, methodology, funding acquisition, writing—review and editing, I.R.D.-G. Both authors have read and agreed to the published version of the manuscript.

Funding: This research was funded by the California Walnut Board and Commission.

Institutional Review Board Statement: Not applicable.

Informed Consent Statement: Not applicable.

Data Availability Statement: The data presented in this study is available in the section "Supplementary Material".

Conflicts of Interest: The authors declare no conflict of interest.

References

1. Miraliakbari, H.; Shahidi, F. Oxidative stability of tree nut oils. *J. Agric. Food Chem.* **2008**, *56*, 4751–4759. [CrossRef] [PubMed]
2. Oliveira, R.; Rodrigues, M.F.; Bernardo-Gil, M.G. Characterization and supercritical carbon dioxide extraction of walnut oil. *J. Am. Oil Chem. Soc.* **2002**, *79*, 225–230. [CrossRef]
3. Maguire, L.S.; O'Sullivan, S.M.; Galvin, K.; O'Connor, T.P.; O'Brien, N.M. Fatty acid profile, tocopherol, squalene and phytosterol content of walnuts, almonds, peanuts, hazelnuts and the macadamia nut. *Int. J. Food Sci. Nutr.* **2004**, *55*, 171–178. [CrossRef] [PubMed]
4. Savage, G.; Dutta, P.; McNeil, D. Fatty acid and tocopherol contents and oxidative stability of walnut oils. *J. Am. Oil Chem. Soc.* **1999**, *76*, 1059–1063. [CrossRef]
5. Madawala, S.; Kochhar, S.; Dutta, P. Lipid components and oxidative status of selected specialty oils. *Int. J. Fats Oils* **2012**, *63*, 143–151. [CrossRef]
6. Greve, L.C.; Labavitch, J.M. Development of rancidity in walnuts. *Walnut Res. Rep. Walnut Mktg. Board Sacram. Calif.* **1985**, *245*, 235.
7. Mahesar, S.A.; Sherazi, S.T.H.; Khaskheli, A.R.; Kandhro, A.A.; Uddin, S. Analytical approaches for the assessment of free fatty acids in oils and fats. *Anal. Methods* **2014**, *6*, 4956–4963. [CrossRef]
8. Kwon, C.W.; Park, K.-M.; Park, J.W.; Lee, J.; Choi, S.J.; Chang, P.-S. Rapid and Sensitive Determination of Lipid Oxidation using the reagent kit based on spectrophotometry (FOODLABfat System). *J. Chem.* **2016**, *2016*, 1–6. [CrossRef]
9. Läubli, M.W.; Bruttel, P.A. Determination of the oxidative stability of fats and oils: Comparison between the active oxygen method (AOCS Cd 12-57) and the Rancimat method. *J. Am. Oil Chem. Soc.* **1986**, *63*, 792–795. [CrossRef]
10. Pawar, N.; Purohit, A.; Gandhi, K.; Arora, S.; Singh, R. Effect of operational parameters on determination of oxidative stability measured by rancimat method. *Int. J. Food Prop.* **2014**, *17*, 2082–2088. [CrossRef]
11. Metrohm. Oxidation Stability of Walnut Oil. Available online: https://www.metrohm.com/en-us/applications/AN-R-013 (accessed on 25 March 2020).
12. Myers, R.H.; Montgomery, D.C.; Anderson-Cook, C.M. *Response Surface Methodology: Process and Product Optimization Using Designed Experiments*; John Wiley & Sons: Hoboken, NJ, USA, 2016.
13. Ferreira, S.C.; Bruns, R.; Ferreira, H.; Matos, G.; David, J.; Brandão, G.; Da Silva, E.; Portugal, L.; Dos Reis, P.; Souza, A.; et al. Box-Behnken design: An alternative for the optimization of analytical methods. *Anal. Chim. Acta* **2007**, *597*, 179–186. [CrossRef] [PubMed]
14. Donis-González, I.R.; Guyer, D.E.; Pease, A. Application of Response Surface Methodology to systematically optimize image quality in computer tomography: A case study using fresh chestnuts (Castanea spp.). *Comput. Electron. Agric.* **2012**, *87*, 94–107. [CrossRef]
15. Rodríguez-González, V.; Femenia, A.; Minjares-Fuentes, R.; González-Laredo, R.F. Functional properties of pasteurized samples of Aloc barbadensis Miller: Optimization using response surface methodology. *Lwt-Food Sci. Technol.* **2012**, *47*, 225–232. [CrossRef]
16. Khir, R.; Pan, Z.; Atungulu, G.G.; Thompson, J.F.; Shao, D. Size and moisture distribution characteristics of walnuts and their components. *Food Bioprocess Technol.* **2013**, *6*, 771–782. [CrossRef]
17. Metrohm. *892 Professional Rancimat Manual M*; Metrohm AG: Herisau, Switzerland, 2017.
18. Farhoosh, R. The effect of operational parameters of the Rancimat method on the determination of the oxidative stability measures and shelf-life prediction of soybean oil. *J. Am. Oil Chem. Soc.* **2007**, *84*, 205–209. [CrossRef]
19. Lenth, R.V. Response-Surface Methods in R, using rsm. *J. Stat. Softw.* **2009**, *32*, 1–17. [CrossRef]
20. Jebe, T.A.; Matlock, M.G.; Sleeter, R.T. Collaborative study of the oil stability index analysis. *J. Am. Oil Chem. Soc.* **1993**, *70*, 1055–1061. [CrossRef]
21. Reynhout, G. The effect of temperature on the induction time of a stabilized oil. *J. Am. Oil Chem. Soc.* **1991**, *68*, 983–984. [CrossRef]

Article

Characterization of Acid-Soluble Collagen from Food Processing By-Products of Snakehead Fish (*Channa striata*)

Thi Mong Thu Truong [1], Van Muoi Nguyen [2], Thanh Truc Tran [2,*] and Thi Minh Thuy Le [1,*]

[1] Department of Fisheries Products Processing, Can Tho University, University Campus II, Can Tho City 94000, Vietnam; ttmthu@ctu.edu.vn
[2] Department of Food Technology, College of Agriculture, Can Tho University, University Campus II, Can Tho City 94000, Vietnam; nvmuoi@ctu.edu.vn
* Correspondence: tttruc@ctu.edu.vn (T.T.T.); ltmthuy@ctu.edu.vn (T.M.T.L.);
Tel.: +84-90-971-2070 (T.T.T.); +84-96-906-2679 (T.M.T.L.)

Abstract: The isolation of acid-soluble collagen (ASC) from by-products of snakehead fish (*Channa striata*), including skin and the mixture of skin and scale, has been investigated. The recovery yield of fish skin ASC (13.6%) was higher than ASC from fish skin and scale (12.09%). Both ASCs were identified as type I collagen and showed maximal solubility at pH 2. Collagen samples from the mixture of skin and scale had higher imino acid content (226 residues/1000 residues) and lower wavenumber in the amide I and amide III region (1642 and 1203 cm^{-1}, respectively) than the fish skin ASC (the imino acid content was 220 residues/1000 residues and the wavenumber in the amide I and amide III were 1663 and 1206 cm^{-1}, respectively. The difference scanning calorimeter (DSC) showed higher thermal stability in ASC from the mixture of skin and scale (T_d of 35.78 °C) than fish skin ASC (34.21 °C). From the result, the denaturation temperature of ASC had a close relationship with the content of imino acid as well as with the degradation of α-helix in amide I and III. These results suggest that collagen could be obtained effectively from snakehead fish by-products and has potential as a realistic alternative to mammalian collagens.

Keywords: acid-soluble collagen; snakehead fish; fish skin; the mixture of skin and scale; denaturation temperature; FTIR

Citation: Truong, T.M.T.; Nguyen, V.M.; Tran, T.T.; Le, T.M.T. Characterization of Acid-Soluble Collagen from Food Processing By-Products of Snakehead Fish (*Channa striata*). *Processes* **2021**, *9*, 1188. https://doi.org/10.3390/pr9071188

Academic Editors: Wei Ma and Hah Young Yoo

Received: 23 May 2021
Accepted: 7 July 2021
Published: 8 July 2021

Publisher's Note: MDPI stays neutral with regard to jurisdictional claims in published maps and institutional affiliations.

1. Introduction

The farming of snakehead fish, especially common snakeheads (*Channa striata*), a popular freshwater fish species, has rapidly spread in Vietnam [1]. Snakehead fish is a popular species and widely consumed in fresh and processed products in the Mekong Delta [2]. Snakehead production has developed rapidly in recent years; total production increased from 5300 tons to 40,000 tons during a 2002–2009 period [1]. The main processed products from snakehead fish in Vietnam contained dried snakehead fish and fish balls. The main steps to produce dried snakehead fish were, removed scales, fillet, remove viscera, and skin. For producing fish balls, the main steps were to fillet and separate the fish meat and the mixture of skin and scale. The by-products discarded during processing mainly consisted of head, skin, bone, fins, scales, and viscera, and accounted for 54% of the snakehead fish weight [3]. The huge amount of these by-products caused serious ecological issues. Furthermore, it was also becoming a potential material for the production of value-added products, especially for the extraction of collagen and gelatin [4].

Collagen has a huge application in food science for forming emulsions, edible films, or using as an agent of antimicrobial, antioxidant, and anti-hypertensive properties during food storage. Furthermore, collagen and gelatin protein hydrolysis could be used as a material for edible film production [5]. The utilization of fish by-products as a material for collagen extraction has been of interest in recent years [6] by the absence of disease transmission and dietary restrictions [7] when compared to the initial sources of collagen extraction as bovine and porcine. Some previous research utilized different fish by-products for collagen extraction, such as the skin of tra catfish, clown knifefish, tilapia [8], golden fish [9], or from the scales of horse mackerel or flying fish [10]. However, the recovery yield of fish collagen is mainly dependent on the kind of fish by-products and fish species involved [11]. Furthermore, some important characteristics of collagen, such as the denaturation temperature, have a close correlation with fish habitat and imino acid content [10,12]. However, we find no information on the collagen extraction from snakehead fish by-products. Therefore, the aim of this study was to isolate acid-soluble collagen (ASC) from snakehead fish skin and the mixture of skin and scale to compare the extracted ASC characteristics and extraction yields between the representative materials for selecting a suitable collagen extraction material.

2. Material and Methods

2.1. Collected and Pretreated Skin and the Mixture of Skin and Scale

Skin and the mixture of skin and scale of snakehead fish were collected from a dried snakehead fish and fish ball company located in An Giang province, Vietnam. The main steps for preparing these by-products from whole snakehead fish were, (i) wash in chilled water, store under ice conditions, and transfer to the laboratory immediately; (ii) cut sample into small pieces and put into PE bags; and (iii) store at $-20\,^{\circ}$C until used.

2.2. Acid Soluble Collagen Extraction (ASC)

ASC was isolated according to the method of [10] with slight adjustments. The snakehead fish skin was soaked and stirred gently in NaOH 0.1 M for 6 h (changing solution every 3 h) at $4\,^{\circ}$C with a material/alkaline solution ratio of 1/8 (*w/v*) to remove the non–collagenous proteins. The mixture of skin and scale were treated to demineralization in EDTA-2Na 0.8 M for 24 h (change solution every 12 h) at $4\,^{\circ}$C at a material/solution ratio of 1/8 (*w/v*). After pretreated, the fish skin and the mixture of skin and scale were washed completely in cold distilled water until a neutral pH was obtained. The collagen extraction of skin and the mixture of skin and scale were conducted by using 10 volumes of acetic acid 0.5 M for 3 days at $4\,^{\circ}$C. The solution obtained after extraction was continuously treated with these main steps: (i) centrifuging at $9000\times g$ for 20 min at $4\,^{\circ}$C to collect the supernatant; (ii) using NaCl for salt-out until obtaining a final concentration of 2.6 M in the presence of tris (hydroxymethyl) aminomethane 0.05 M at pH 7.0; (iii) centrifuging again at $9000\times g$ for 20 min at $4\,^{\circ}$C to collect the precipitate; (iv) dissolving in 0.5 M acetic acid, dialyzing and lyophilizing at $-20\,^{\circ}$C for 12 h; then (v) collecting the ASC by using a freeze-dryer.

2.3. Yield of Extracted ASC

Basing on the hydroxyproline content (Hyp) in extracted ASC and the crude material, the extraction yield of ASC was calculated following the described of [13] as below:

$$\text{Yield (\%)} = \frac{\text{Hyp content in extracted ASC (mg/L)} \times \text{Total extracted volume (L)}}{\text{Hyp content in crude material (mg/g)} \times \text{dry weight of crude material (g)}}$$

2.4. Viscosity of ASC

The collagen sample was dissolved completely in 0.1 M acetic acid to obtain a solution with a final concentration of 6.67%. The viscosity of the collagen solution was measured as described by [14] by using a digital viscometer (Brookfield DV, RVDV-11+CP, Middleboro, MA 02346, USA) with the speed of the spindle at 100 rpm.

2.5. The Solubility of Collagen at Different pHs and Various NaCl Concentrations

For determining the changing ASC solubility at different pH, a collagen sample at a final concentration of 3 mg/mL was prepared by dissolving in 0.1 M acetic acid and separated into 10 parts, the volume of each part was 8 mL. We adjusted pH of each part by using either 6 M HCl or 6 M NaOH from 1 to 10 and then volume up to 10 mL by adding distilled water. Each 10 mL of collecting solution was stirred for 30 min at 4 °C and continuously centrifuged at 9000× *g* at 4 °C for 20 min. The resulting supernatant was collected, and protein content was checked using Lowry's method [15] with bovine serum albumin as a protein standard. The solubility of the supernatant at each pH was calculated by comparing it with the pH showing the highest solubility by the equation below:

$$\text{The solubility at each pH (\%)} = \frac{\text{The protein content at each pH}}{\text{The highest protein content}} \times 100$$

For testing the changing of ASC solubility at various NaCl concentrations, the collagen solution at the concentration of 6 mg/mL was obtained by dissolving collagen in 0.1 M acetic acid. Collagen solution was separated into 6 parts, with the volume of each part being 5 mL. Mixing each part with 5 mL of NaCl solution at 0.2, 0.4, 0.6, 0.8, 1, and 1.2 M while stirring for 30 min at 4 °C, and then centrifuging at 8000× *g* at 4 °C for 20 min. The resulting supernatant was collected and checked protein content as described as above. The supernatant obtained at each NaCl concentration was checked solubility and compared with the NaCl concentration showing the highest solubility.

2.6. Amino Acid Analysis

The amino acid composition of ASC from the snakehead fish by-products was analyzed according to the method of [10]. Approximately 0.1 g of ASC was measured into a media bottle for each analysis. Then we added 5 mL of oxidation solution into the bottle while continuously stirring and placed it into a refrigerator at 0 °C for 16 h. Subsequently, these ASC samples were combined with 0.84 g of sodium disulfite and 25 mL of hydrolysis solution. After that, the ASC samples were hydrolyzed by furnacing at 110 °C for 23 h. The hydrolytes were derivatized, using sample diluents for diluting, and compared with the analyzed standard of amino acids. was used for Amino acids analysis was conducted with 20 μL of the sample by using the Amino Acid Analyzer system (Biochrom 32+, Holliston, MA 01746, USA). The data were expressed in amino acid residues/1000 residues of amino acid contents.

2.7. The Molecular Weight of Collagen by SDS-PAGE

The protein profiles of ASC were accomplished referring to [16], as described by [8], with slight adjustments. ASC samples dissolved in sample buffer contained Tris-HCl 0.5 M, pH 6.8, including SDS 10% (*w/v*) and glycerol 20% (*v/v*) in the presence of mercaptoethanol 10% (*v/v*) at a ratio of collagen/sample buffer 1/2 (*v/v*); then they were loaded onto polyacrylamide gel 7.5% (Biorad, Hercules, CA 94547, USA) with protein molecular weight markers (Sigma Chemical Co., St. Louis, MO, USA). After being separated, gels were stained and destained. By comparing with protein markers, the molecular weight of protein bands was estimated.

2.8. FTIR Spectra of Collagen

FTIR spectroscopy of collagen samples was collected using a PerkinElmer MIR/NIR Frontier spectrometer in MIR mode. The data were analyzed using the software program SPECTRUM (PerkinElmer, Hopkinton, MA 01748, USA). The recorded spectra were obtained from a PerkinElmer MIR/NIR Frontier spectrometer, with spectra wavenumbers from 4000 to 400 cm^{-1}.

2.9. Thermal Stabilization of ASC by DSC (Differential Scanning Calorimeter)

The sample was conducted by dissolving ASC in 0.1 M acetic acid at a solid/acetic acid ratio of 1:40 (*w/v*). The thermal stability was performed by DSC (Pyris 1; PerkinElmer Co., Ltd., Yokohama, Japan) with a rate of scanning of 1 °C/min in the temperature range of 20–50 °C as described by [8]. By estimating the endothermic peak of the DSC

Thermogram, the total denaturation enthalpy (ΔH) and the denaturation temperature (T_d) were determined.

2.10. Data Analysis

All of the experiments were carried out in triplicate. The data was shown as a standard deviation of the mean (S.D.M). Using Duncan's tests we determined the variable differences and analyses of data were performed in SPSS 16.0 software.

2.11. Animal Welfare

The research was carried out according to the project CT2020.01.TCT.03, approved by the Director of the Department of Science, Technology, and Environment, under the Minister of Education and Training of Vietnam on 9 January 2020.

All experiments were carried out in accordance with national guidelines on the protection of animals and experimental animal welfare in Vietnam following the Law on Animal Health, 2015, Vietnam National Assembly, No. 79/2015/QH13, approved 19 June 2015.

3. Results and Discussion

3.1. Extraction Yield and Viscosity of ASC According to Snakehead Fish Used Part

The extraction yields and viscosity of ASC from the skin and the mixture of skin and scale of snakehead fish are shown in Table 1. The ASC yield in snakehead fish skin (13.6%) was higher than the yield of ASC from the mixture of skin and scale (12.09%). However, these extraction yields from by-products of snakehead fish (12.09–13.6%) showed a similar tendency of collagen from the skin of tra catfish (13.81%) and tilapia (12.5%), but was lower than the yield of ASC from clown knifefish skin (16.04%) [8] and black carp skin (15.5%) [17]. Thus, Duan et al. [18] reported that the ASC extraction yield was dependent on the fish skin by the differences in skin matrices and the dispensation of alignment in the constituent of skin.

Table 1. The extraction yield (%) and viscosity (mPa·s) of ASC from by-products of snakehead fish.

Sample	Recovery Yield (%)	Viscosity (mPa·s)
Skin collagen	13.60 ± 0.13 [a]	37.5 ± 1.30 [a]
Skin and scale collagen	12.09 ± 0.20 [b]	21.3 ± 0.95 [b]

Data are expressed as mean ± standard deviation ($n = 3$). Different superscripts in the same column indicate statistical differences ($p < 0.05$).

The viscosity of skin collagen was higher than 1.76 times when compared to collagen from the mixture of skin and scale. It might be explained that the skin matrices were loose in comparison to the skin and scale mixture and showed a higher solubility in the solution of acetic acid during the extraction process than the skin and scale mixture. This leads to a higher extraction yield and higher viscosity of ASC from snakehead fish skin.

3.2. Effect of Various pH and NaCl Concentrations on ASC Solubility

The solubility alteration of ASC from skin and the mixture of skin and scales at a pH range from 1 to 10 are presented in Figure 1. The high solubility in an acidic pH range, and the solubility decline in the alkaline pHs region by ASC precipitating when pH closed to pI [19], could be observed in both ASCs. Collagen samples showed maximal solubility at pH two. The solubility of both collagen samples showed a slight increase at pH above nine because of the increased repulsion of collagen chains, as the pH gain to a higher value than the pI [20]. This similar tendency in the changing of ASC solubility in this research was agreeable to the reports of [9,18] about the solubility of skin ASC from golden carp and three freshwater fish (tra catfish, clown knife fish, and tilapia), respectively.

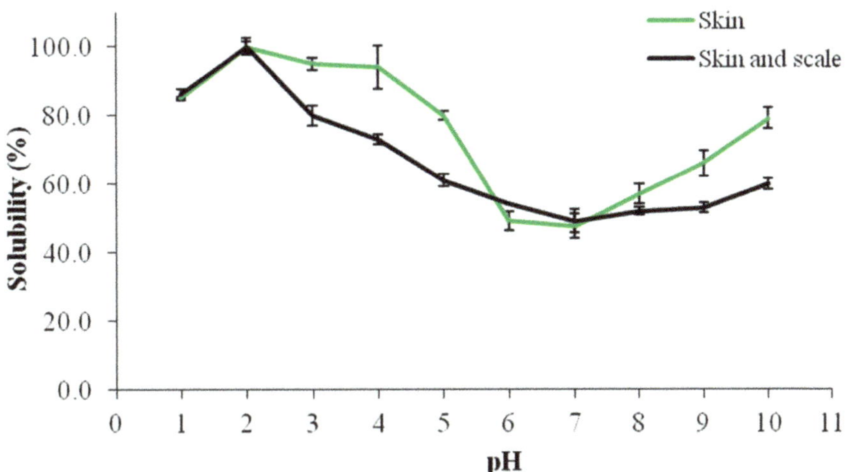

Figure 1. The solubility of ASC from by-products of snakehead fish at different pH values.

NaCl concentration also had a large effect on the ASC solubility. Figure 2 showed the solubility of both ASCs at various NaCl concentrations. The similar solubility behaviors could be observed in both collagen samples, with high solubility at the range of NaCl concentrations from 0.2 to 0.4 M (92.16–100%). The dramatic decline in the solubility was shown with NaCl concentrations above 0.4 M because the precipitation of protein by the hydrophobic–hydrophobic interactions was increased [21].

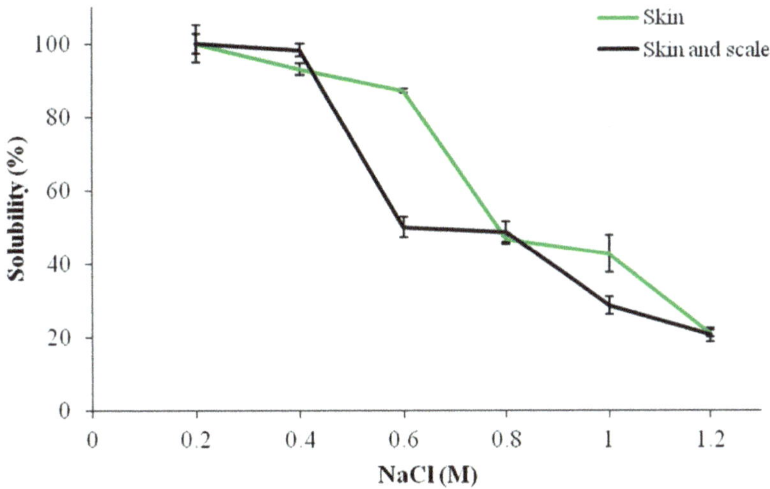

Figure 2. The solubility of ASC from by-products of snakehead fish at different NaCl concentrations.

3.3. Amino Acid Composition of ASC from Snakehead Fish

Amino acid compositions from the ASC of skin and the mixture of snakehead fish skin and scales are presented in Table 2.

Table 2. Amino acid content of collagen from skin and skin-scale of snakehead fish.

Amino Acid	Amino Acid Content (Residues/1000 Residues)	
	ASC from Skin	ASC from Skin and Scale
Aspartic acid	54 ± 3	49 ± 2
Threonine	22 ± 2	21 ± 2
Serine	35 ± 2	36 ± 3
Glutamic acid	74 ± 4	76 ± 4
Glycine	307 ± 7	314 ± 9
Alanine	89 ± 5	91 ± 6
Valine	26 ± 1	23 ± 1
Cystein	2 ± 1	2 ± 1
Methionine	12 ± 1	12 ± 2
Tryptophan	0	0
Isoleucine	9 ± 2	9 ± 2
Leucine	28 ± 3	24 ± 3
Tyrosine	5 ± 1	4 ± 1
Phenylalanine	18 ± 2	17 ± 2
Hydrolysine	6 ± 1	6 ± 1
Lysine	31 ± 2	30 ± 3
Histidine	6 ± 1	6 ± 1
Arginine	56 ± 3	54 ± 2
Hydroxyproline	94 ± 5	95 ± 4
Proline	126 ± 5	131 ± 6
Imino acid (Proline and Hydroxyproline)	220 ± 6	226 ± 7

The amino acid compositions of ASC from snakehead fish by-products, expressed as residues/1000 residues, showed a similar tendency. It is well known that collagen type I is abundant in glycine, which accounted for 307 and 314 residues/1000 residues, from ASC of fish skin and ASC of skin and scale mixture. The α-chains in the triple-helix structure of collagen had the general formula Glycine-Proline-Hydroxyproline [22]. The total amount of proline and hydroxyproline (imino acid) of ASC from the skin (220 residues) and the mixture of skin and scale (226 residues) was higher than the ASC from tra catfish, tilapia, and clown knifefish skin (192, 195, and 200/1000 residues, respectively) [8] as well as the skin collagen of Pacific cod (157 residues, [23]) and skin of bighead carp (165 residues, [24]). Normally, the imino acids contribute to the stabilization of collagen via sustaining the wholeness of the triple helix structure. Why the content of imino acid has been different among collagens from various fish species is most likely due to the temperature in the fish habitats [25]. Higher imino acid content of ASC is found in fish species living in warm compared to cold water fishes [10].

3.4. Protein Pattern of ASC from Snakehead Fish

SDS-PAGE patterns of collagen from fish skins and the mixture of skin and scale are shown in Figure 3.

SDS-PAGE profiles showed a similar pattern in the protein of ASC from skin and mixture of skin and scale. All ASCs included α1, α2, as well as β chains as major components; the low band intensity of the γ chain was found at a very low content, and was identified as a type I collagen. Previous reports have classified ASCs from fish by-products such as skin and scale as collagen type I, including the skin of tilapia, clown knifefish, and Pacific cod [8,23], as well as in the scales of horse mackerel and flying fish [10].

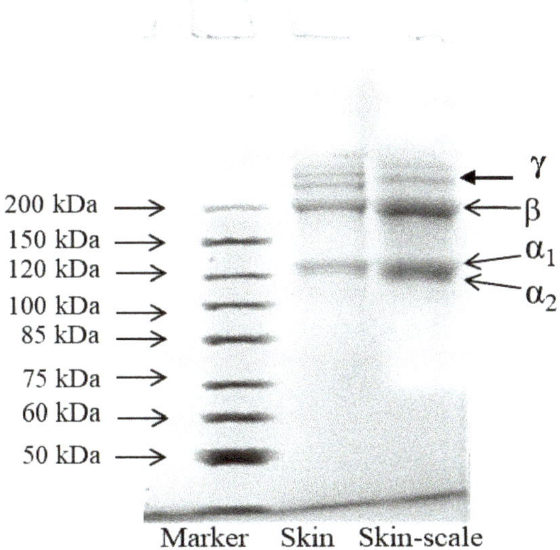

Figure 3. Protein patterns of collagen from skin and the mixture of skin and scale of snakehead fish.

3.5. FTIR Spectra of ASC from Snakehead Fish

The FTIR spectra of skin and the mixture of skin and scale ASCs from snakehead fish are shown in Figure 4. Both samples had amide A (3400–3440 cm^{-1}), amide I (1600 to 1700 cm^{-1}), amide II (1500 to 1600 cm^{-1}), and amide III (1200 to 1300 cm^{-1}) bands, which were in agreement with the research of [9]. Amide A of both ASCs was observed at a wavenumber of 3422 and 3414, respectively. Furthermore, the spectra wavenumber included amide I (1663 and 1642 cm^{-1} in ASC from skin and the mixture of skin and scale, respectively), amide II (1564 and 1552 cm^{-1}), and amide III (1206 and 1203 cm^{-1}). Amide A, normally associated with N–H stretching vibrations coupled with hydrogen bonds, appears in the spectra range of 3400–3440 cm^{-1}. Amide I bands are associated with C=O stretching vibrations in peptides, with the main function to form a secondary protein structure. Amide II (~1500 cm^{-1}) represents N–H bending coupled to C–N stretching. The triple-helix structure of collagen is involved with amide III [20]. The absorption bands of amide I and amide III from the skin ASC were higher than the mixture of skin and scale ASC, indicating an association between the transition of α-helix by uncoupling of intermolecular cross-links and the interruption of intramolecular bonding [26]. Moreover, it may lead to the thermal stability of ASC from skin and scale mixture being higher than that from skin collagen. The correlation between denaturation temperature of ASC and the decline of wavenumber of amide I and amide III bands has been described in the study by Thuy et al. [8], who reported that ASC from the skin of clown knifefish, with the lowest wavenumber of amide I and amide III, showed the highest thermal stability in comparison to collagen from tra catfish and tilapia skin.

3.6. Thermal Properties of ASC from Snakehead Fish

The temperature of collagen denaturation from snakehead fish skin and the mixture of skin and scale are presented in Figure 5. The endothermic peak of skin ASC was observed with T_d of 34.21 °C, which was slightly lower than that of ASC from the mixture of skin and scale (35.78 °C). Some previous research reported the difference of T_d from different collagen sources and suggested that T_d depended on fish species, fish tissues used for extraction, age, habitat temperature and environments, and seasons [10,27–29]. The denaturation temperature of collagen had a close relationship with the imino acid content [10,12]. Furthermore, Thuy et al. [8] reported that the denaturation temperature of

ASC might not only depend on the imino acid content but is also directly related to the degradation of wavenumber in the amide I and amide III region. ASC from skin and scale mixture with higher content of imino acid (226 residues/1000 residues of amino acid) and a lower wavenumber in amide I and amide III was shown with the T_d higher than that of skin collagen (220 residues).

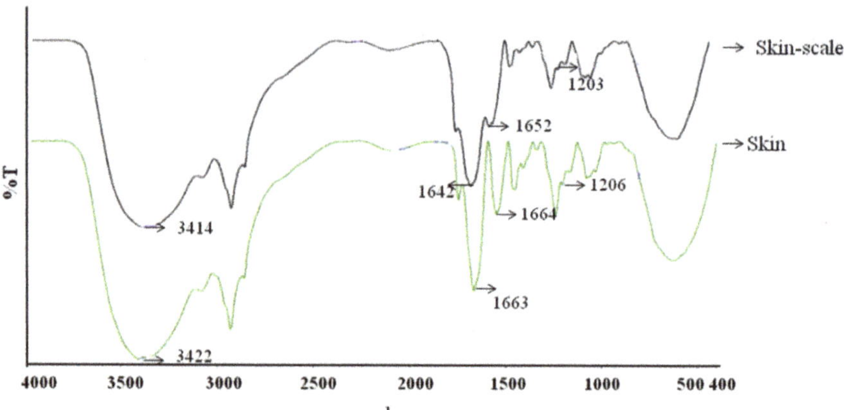

Figure 4. FTIR of collagen from skin and skin-scale mixture of snakehead fish.

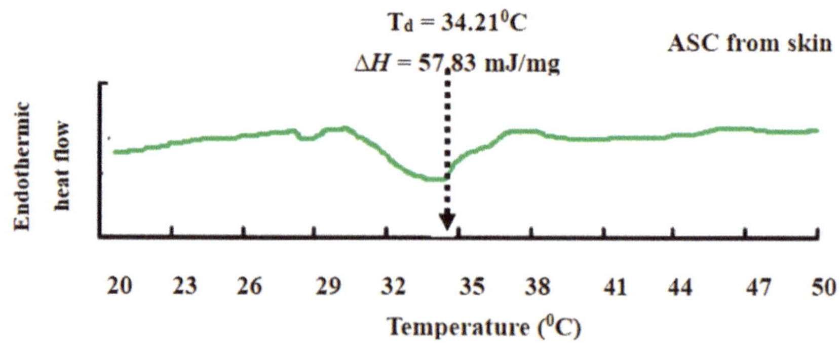

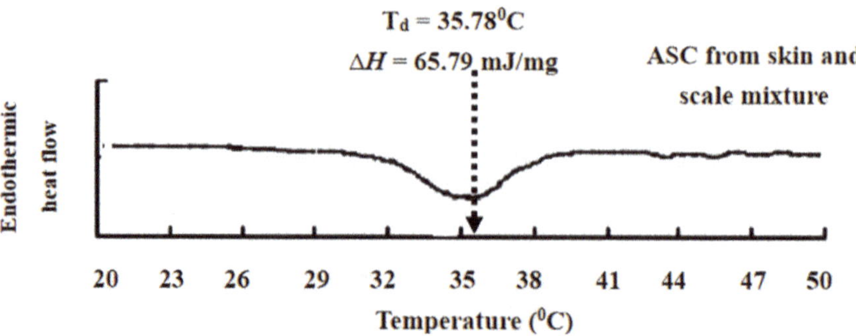

Figure 5. Denaturation temperature of collagen from by-products of snakehead fish.

4. Conclusions

Based on SDS-PAGE, both ASC from snakehead fish skin and the mixture of skin and scale have been isolated and classified as type I collagen. ASC from fish skin showed the recovery yield was higher than those from the mixture of fish skin and scale. However, collagen samples from skin and scale had higher thermal stability than skin collagen. Thus, the by-products from snakehead fish could be an alternative source for collagen extraction to reduce environmental pollution and increase the value of fish processing by-products. Collagen from snakehead fish by-products could be applied in food science, especially in food packaging.

Author Contributions: T.M.T.L., V.M.N., and T.T.T. designed this study and interpreted the results. T.M.T.T. collected test data and drafted the manuscript. All authors have read and agreed to the published version of the manuscript.

Funding: This research was funded by Ministry of Education and Training, Vietnam, grant number CT2020.01.TCT.03.

Institutional Review Board Statement: The study was conducted according to the national guidelines on the protection of animals and experimental animal welfare in Vietnam following the Law on Animal Health, 2015, Vietnam National Assembly, No. 79/2015/QH13, approved 19 June 2015.

Informed Consent Statement: Not applicable.

Acknowledgments: This research cost was supported by the CT2020.01 program (CT2020.01.TCT.03 project) funding from the Ministry of Education and Training, Vietnam.

Conflicts of Interest: None of the authors have any conflicts of interest.

References

1. Sinh, L.X.; Navy, H.; Pomeroy, R.S. Value chain of snakehead fish in the lower Mekong basin of Cambodia and Vietnam. *Aquac. Econ. Manag.* **2014**, *18*, 76–96. [CrossRef]
2. Navy, H.; Truong, H.M.; Pomeroy, R. The impacts of climate change on snakehead fish value chains in the lower Mekong basin of Cambodia and Vietnam. *Aquac. Econ. Manag.* **2017**, *21*, 261–282. [CrossRef]
3. Muoi, N.V.; Truc, T.T.; Ngan, V.H. The influence of additives on frozen snakehead fish surimi and the application of transglutaminase to fish cakes. *Acta Sci. Pol. Technol. Aliment.* **2019**, *18*, 125–133. [CrossRef]
4. Elavarasan, K.; Kumar, A.; Uchoi, D.; Tejpal, C.S.; Ninan, G.; Zynudheen, A.A. Extraction and characterization of gelatin from the head waste of tiger tooth croaker (*Otolithes ruber*). *Waste Biomass Valoriz.* **2016**, *8*, 851–858. [CrossRef]
5. Jancikova, S.; Jamróz, E.; Kulawik, P.; Tkaczewska, J.; Dordevic, D. Furcellaran/gelatin hydrolysate/rosemary extract composite films as active and intelligent packaging materials. *Int. J. Biol. Macromol.* **2019**, *131*, 19–28. [CrossRef] [PubMed]
6. Kittiphattanabawon, P.; Benjakul, S.; Visessanguan, W.; Nagai, T.; Tanaka, M. Characterisation of acid-soluble collagen from skin and bone of bigeye snapper (*Priacanthus tayenus*). *Food Chem.* **2005**, *89*, 363–372. [CrossRef]
7. Ahmad, M.; Nirmal, N.P.; Chuprom, J. Molecular characteristics of collagen extracted from the starry triggerfish skin and its potential in the development of biodegradable packaging film. *RSC Adv.* **2016**, *6*, 33868–33879. [CrossRef]
8. Thuy, L.M.T.; Muoi, N.V.; Truc, T.T.; Takahashi, K.; Osako, K. Comparison of acid-soluble collagen characteristic from three important freshwater fish skins in Mekong Delta Region, Vietnam. *J. Food Biochem.* **2020**, *44*, e13397.
9. Ali, A.M.M.; Benjakul, S.; Prodpran, T.; Kishimura, H. Extraction and Characterisation of Collagen from the Skin of Golden Carp (*Probarbus jullieni*), a Processing By-Product. *Waste Biomass Valoriz.* **2018**, *9*, 783–791. [CrossRef]
10. Thuy, L.T.M.; Okazaki, E.; Osako, K. Isolation and characterization of acid-soluble collagen from the scales of marine fishes from Japan and Vietnam. *Food Chem.* **2014**, *149*, 264–270. [CrossRef] [PubMed]
11. Nagai, T.; Suzuki, N. Isolation of collagen from fish waste material—Skin, bone and fins. *Food Chem.* **2000**, *68*, 277–281. [CrossRef]
12. Bae, I.; Osatomi, K.; Yoshida, A.; Osako, K.; Yamaguchi, A.; Hara, K. Biochemical properties of acid-soluble collagen extracted from the skins of underutilised fishes. *Food Chem.* **2008**, *108*, 49–54. [CrossRef]
13. Wang, L.; An, X.; Yang, F.; Xin, Z.; Zhao, L.; Hu, Q. Isolation and characterisation of collagen from the skin, scale and bone of deep-sea redfish (*Sebastes mentella*). *Food Chem.* **2008**, *108*, 616–623. [CrossRef]
14. Jamilah, B.; Tan, K.W.; Umi Hartina, M.R.; Azizah, A. Gelatins from three cultured freshwater fish skins obtained by liming process. *Food Hydrocoll.* **2011**, *25*, 1256–1260. [CrossRef]
15. Lowry, O.H.; Rosebrough, N.J.; Farr, A.L.; Randall, R.J. Protein measurement with Folin phenol reagent. *J. Biol. Chem.* **1951**, *193*, 256–275. [CrossRef]
16. Laemmli, U.K. Cleavage of structural protein during assembly head of bacteriophage T4. *Nature* **1970**, *277*, 680–685. [CrossRef] [PubMed]

17. Jia, Y.; Wang, H.; Li, Y.; Wang, M.; Zhou, J. Biochemical properties of skin collagens isolated from black carp (*Mylopharyngodon piceus*). *Food Sci. Biotechnol.* **2012**, *21*, 1585–1592. [CrossRef]
18. Duan, R.; Zhang, J.; Du, X.; Yao, X.; Konno, K. Properties of collagen from skin, scale and bone of carp (*Cyprinus carpio*). *Food Chem.* **2009**, *112*, 702–706. [CrossRef]
19. Jongjareonrak, A.; Benjakul, S.; Visessaguan, W.; Tanaka, M. Isolation and characterisation of collagen from bigeye snapper (*Priacanthus macracanthus*) skin. *J. Sci. Food Agric.* **2005**, *85*, 1203–1210. [CrossRef]
20. Muyonga, J.H.; Cole, C.G.B.; Duodu, K.G. Characterization of acidic soluble collagen from skins of young and adult Nile Perch (*Lates niloticus*). *Food Chem.* **2004**, *85*, 81–89. [CrossRef]
21. Damodaran, S. Amino Acids, Peptides and Proteins. In *Food Chemistry*; Fennema, O.R., Ed.; Marcel Dekker: New York, NY, USA, 1996; pp. 321–429.
22. Regenstein, J.M.; Zhou, P. Collagen and Gelatin from Marine by-Products. In *Maximising the Value of Marine By-Products*; Shahidi, F., Ed.; Woodhead Publishing Limited: Cambridge, UK, 2007; pp. 279–303.
23. Liu, D.; Liang, L.; Regenstein, J.M.; Zhou, P. Extraction and characterisation of pepsin-solubilised collagen from fins, scales, skins, bones and swim bladders of bighead carp (*Hypophthalmichthys nobilis*). *Food Chem.* **2012**, *133*, 1441–1448. [CrossRef]
24. Sun, L.; Li, B.; Song, W.; Si, L.; Hou, H. Characterization of Pacific cod (*Gadus macrocephalus*) skin collagen and fabrication of collagen sponge as a good biocompatible biomedical material. *Process. Biochem.* **2017**, *63*, 229–235. [CrossRef]
25. Foegeding, E.A.; Lanier, T.C.; Hultin, H.O. Characteristics of Edible Muscle Tissues. In *Food Chemistry*; Fennema, O.R., Ed.; Marcel Dekker, Inc.: New York, NY, USA, 1996; pp. 879–942.
26. Kittiphattanabawon, P.; Benjakul, S.; Visessanguan, W.; Shahidi, F. Effect of extraction temperature on functional properties and antioxidative activities of gelatin from shark skin. *Food Bioprocess Technol.* **2012**, *5*, 2646–2654. [CrossRef]
27. Chuaychan, S.; Benjakul, S.; Kishimura, H. Characteristics of acid- and pepsin-soluble collagens from scale of seabass (*Lates calcarifer*). *LWT Food Sci. Technol.* **2015**, *63*, 71–76. [CrossRef]
28. Tang, L.; Chen, S.; Su, W.; Weng, W.; Osako, K.; Tanaka, M. Physicochemical properties and filmforming ability of fish skin collagen extracted from different freshwater species. *Process. Biochem.* **2015**, *50*, 148–155. [CrossRef]
29. Liu, Y.; Ma, D.; Wang, Y.; Qin, W. A comparative study of the properties and self-aggregation behavior of collagens from the scales and skin of grass carp (*Ctenopharyngodon idella*). *Int. J. Biol. Macromol.* **2018**, *106*, 516–522. [CrossRef] [PubMed]

Review

A Review on the Extraction and Processing of Natural Source-Derived Proteins through Eco-Innovative Approaches

Giselle Franca-Oliveira, Tiziana Fornari and Blanca Hernández-Ledesma *

Institute of Food Science Research (CIAL, CSIC-UAM, CEI UAM+CSIC), Nicolás Cabrera, 28049 Madrid, Spain; gisellefrancaoliveira@gmail.com (G.F.-O.); tiziana.fornari@uam.es (T.F.)
* Correspondence: b.hernandez@csic.es; Tel.: +34910017970

Abstract: In addition to their nutritional and physiological role, proteins are recognized as the major compounds responsible for the rheological properties of food products and their stability during manufacture and storage. Furthermore, proteins have been shown to be source of bioactive peptides able to exert beneficial effects on human health. In recent years, scholarly interest has focused on the incorporation of high-quality proteins into the diet. This fact, together with the new trends of consumers directed to avoid the intake of animal proteins, has boosted the search for novel and sustainable protein sources and the development of suitable, cost-affordable, and environmentally friendly technologies to extract high concentrations of valuable proteins incorporated into food products and supplements. In this review, current data on emergent and promising methodologies applied for the extraction of proteins from natural sources are summarized. Moreover, the advantages and disadvantages of these novel methods, compared with conventional methods, are detailed. Additionally, this work describes the combination of these technologies with the enzymatic hydrolysis of extracted proteins as a powerful strategy for releasing bioactive peptides.

Keywords: food proteins; novel extraction methodologies; enzymatic hydrolysis; bioactive peptides

Citation: Franca-Oliveira, G.; Fornari, T.; Hernández-Ledesma, B. A Review on the Extraction and Processing of Natural Source-Derived Proteins through Eco-Innovative Approaches. *Processes* **2021**, *9*, 1626. https://doi.org/10.3390/pr9091626

Received: 29 July 2021
Accepted: 3 September 2021
Published: 9 September 2021

Publisher's Note: MDPI stays neutral with regard to jurisdictional claims in published maps and institutional affiliations.

1. Introduction

Proteins are essential macronutrients involved in the growth and development of the body. In addition to their nutritional and physiological properties, the techno-functional characteristics of proteins are responsible for the appearance, texture, and stability of food products. Moreover, proteins have been demonstrated to be a source of peptides capable of exerting multiple biological activities after their release by hydrolysis, gastrointestinal digestion, and/or food processing [1].

The incorporation of high-quality proteins into the everyday diet is a prevalent theme in current research. To meet consumer trends in limiting the intake of animal proteins, nutritionists and food industries are exploring the use of novel and sustainable protein sources from plants, insects, and algae. However, satisfying this demand requires the simultaneous development of suitable, cost-effective, and eco-friendly technologies to extract higher concentrations of valuable proteins to be incorporated into food and supplements [2]. The application of conventional methods generally results in lower extraction yields due to protein degradation from extreme pH, temperature, solvent conditions, and long extraction times. Therefore, researchers are currently focused on non-thermal green technologies for improving extraction efficiency and reducing protein degradation. These methods generally have no damaging effect on the environment. Moreover, the minimal use of toxic chemicals and reactants makes extracted proteins safe for animal and human consumption. Other than improving the protein yield, these innovative techniques can enhance their nutritional and techno-functional properties as well as their potential as a source of bioactive peptides. Understanding the principles of each of these new methodologies, and their advantages and disadvantages in comparison with conventional methods, is essential to advance in the knowledge of their applications in the food industry, and to achieve extractions, from new sources, of high-purity proteins with interesting properties at high concentrations. Thus, this review aims to respond to this need by providing all the existing evidence on the emergent extraction technologies applied on natural protein sources. Moreover, the application of some novel technologies, in combination with the enzymatic hydrolysis of extracted proteins with the aim to release bioactive peptides, is also described.

2. Extraction of Proteins from Natural Sources

2.1. Chemical Extraction Techniques

The classification of chemical methods is based on the solvent used, such as water, alkali, organic solvents, and acids. Although the efficiency of the methodology primarily depends on the nature of the protein sample, the processing conditions have also been demonstrated to show their influence on protein recovery [3]. Aqueous extraction is a frequently used method due to the high protein solubility and stability of isolated proteins in water. Moreover, this methodology shows other advantages such as its easy operation conditions and low cost [4]. Generally, extraction with water is performed under basic conditions, as it has been described for proteins from different plant sources such as mung bean (*Vigna radiata*) [5], grass pea (*Lathyrus sativus*) [6], rice and rice bran [7,8] and tomato [9,10], among others. However, extreme extraction conditions, such as high temperature or high alkaline conditions, could influence the thermal, conformational, and functional properties of protein fractions, reducing their nutritional value and degrading their bioactive compounds [11]. Mild acidic conditions have also been reported to efficiently extract proteins from sunflower with a 23–26% rate recovery [12].

To extract proteins containing non-polar, hydrophobic and/or aromatic amino acid residues, organic solvents like ethanol, butanol, and acetone are required [13]. Thus, a recent study has described the application of an ethanol-petroleum ether combination in water to extract proteins from *Moringa olifera* seeds, reaching 33% of recovery after the purification of the protein [4].

2.1.1. Aqueous Two-Phase System (ATPS)

Currently, aqueous two-phase system (ATPS) is being used for efficient protein extraction due to the associated properties, such as the hydrophobicity of the phase system, the electrical potential between phases, molecular size, and the bioaffinity of the protein [2]. ATPS is a multifunctional technique that allows separating, concentrating, and purifying proteins. It is based on the mixture of two components of different natures. The appropriate choice of these two components guarantees the completion of two defined and equilibrated layers. The benefits of the ATPS technique are of great interest to the scientific community. Its attributes of rapidity, flexibility, economical convenience, and biocompatibility ensure a higher selectivity, purity, and extraction yield than with conventional systems. On top of that, phase constituents do not denature proteins; they can stabilize protein structures within their biological activity [14]. A recent review summarizing the existing evidence on the application of ATPS for the recovery of valuables, as well as the elimination of contaminants from industrial waste discharges, has been published [15]. Among these high-value components, proteins have been demonstrated to be efficiently extracted by ATPS. Therefore, two-phase systems constituted by sodium citrate and polyethylene glycol, or sodium citrate and ethanol, have been recently reported as useful in extracting proteins from shrimp (*Litopenaeus vannamei*) waste and microalgae (*Arthrospira platensis*) [16,17].

2.1.2. Subcritical Water Extraction (SWE)

Subcritical water (SW) extraction (SWE) is a technique based on the use of hot water in the range from water's normal boiling point (100 °C) to water's critical temperature (374 °C), while using a high pressure to maintain water in its liquid state within those temperatures (usually 220–230 bar). The increase in temperature significantly modifies water's properties. For example, its viscosity and density decrease, but its compressibility remains low, and indeed the most important change is the temperature-dependent decrease of water's dielectric constant. Thus, the hydrogen bonding structure of water is weakened, which enables the solubilization of moderately polar and nonpolar compounds. Furthermore, SW produces a high-ion product, a property suitable for hydrolysis reactions and thus, when compared with water at lower temperatures and ambient pressures, SW conditions enhance the depolymerization of polysaccharides and the generation of smaller soluble protein fractions [18].

Mlyuka et al. [19] presented SWE as a strategic alternative for the food industry with promising potential in the selective extraction of bioactive compounds and in green hydrolysis reactions, while stressing scaling as the major challenge facing its commercial use. SWE is an efficient, cheap, fast, and environmentally friendly technology. Several reviews [19–21] analyzed the tuning of operation conditions in different types of raw materials, to maximize extraction yield and/or to avoid the degradation and decomposition of desired products. In general, in comparison with conventional alkali or enzymatic hydrolysis, SW provides comparable or higher yields in shorter times. In a very recent contribution, Álvarez-Viñas et al. (2021) reviewed the most important features in the SWE/hydrolysis of proteins, the different modes of operation, and analyzed the effect of process conditions on the product properties [22]. The authors highlighted the necessity of establishing optimal conditions as a compromised solution for processing proteins derived from agro-food wastes and algal biomass, suggesting that could be desirable in the stage wise operation to sequentially obtain high-valued fractions. Nevertheless, the formation of allergenic and/or toxic peptides from wastes and biomasses during protein extraction and hydrolysis should be prudently tested.

2.2. Enzyme-Assisted Extraction

Enzyme-assisted extraction (EAE) is a green technology based, firstly, on the action of degrading enzymes within the major components of the cell wall, such as cellulose, hemicellulose, and/or pectins, resulting in the disruption of the wall and the release of cellular proteins [23]. Secondly, proteases break down the high molecular weight cell proteins into smaller and more soluble portions, thus providing valuable extraction conditions [2]. Although EAE has been characterized by its long processing time, high costs, elevated energy consumption, and irreversible carbohydrate-protein matrix disruption, it has become an emergent strategy showing advantages in comparison with conventional solvent-based extraction methods [23]. Moreover, products obtained using this technology evidence a higher purity and suitability for human consumption [24–26]. Thus, Sari et al. demonstrated the higher protein yield resulting from the extraction of proteins from soybean (*Glycine max*) and rapeseed (*Brassica napus subsp. napus*) meals when serine, endo, and exoproteases were used in comparison to the protein yield obtained without enzyme addition [24]. Similarly, Rommi et al. [27] demonstrated the beneficial effects of pectinolytic enzymes acting on pectic polysaccharides and glucans on the extraction yield of proteins from rapeseed press cakes made from cold oil processing. The use of enzymes increased the protein yield by 1.7 times in comparison with that obtained without enzymes. In addition, these authors found that the enzymatic hydrolysis of carbohydrates at a pH of 6 allowed for the extraction of rapeseed press cake proteins with a higher solubility and dispersion stability than those obtained using an alkaline extraction that provoked their partial denaturation [28]. Chirinos et al. also found that the EAE (alcalase) of proteins from sacha inchi (*Plukenetia volubilis* L.) kernel meal resulted in a higher ($\approx$ 1.5-fold) protein recovery than that obtained through alkaline extraction [29]. More recently, bi-(α-amylase and amyloglucosidase) and tri-enzyme (α-amylase, amyloglucosidase, and β-1,3,4-glucanase) treatments were applied to extract proteins from defatted barley flour, obtaining a protein yield of 49% and 78.3%, respectively [30]. Similarly, defatted soybean flour was treated with xylanase, pectinase, cellulase, and a cocktail of commercial carbohydrases within alkaline extraction, resulting in the increase of the protein yield by 21% compared to the 2 h alkaline extraction without enzymatic treatment [31]. In a recent study, four enzyme preparations, followed by assisted alkaline extraction, were tested for protein extraction from the seaweed *Palmaria palmata* [32]. These authors found that the most efficient treatments were the combination of Celluclast® 0.2% *w/w* plus Alcalase® 0.2% *w/w* (90.0% extraction efficiency), or Shearzyme® 0.2% *w/w* plus Alcalase® 0.2% *w/w* (85.5% extraction efficiency). Moreover, these methods allowed for improving the amino acid profile, the essential amino acid score, and the ratio of the extracted proteins, in which their potential as source of bioactive peptides was also improved, making this methodology suitable for extracting proteins with interesting nutritional and functional properties. Recent studies have reported on the suitability of enzyme-assisted extraction to recover proteins from other sources, such as sugar beet (*Beta vulgaris* L.) leaves [33] and almond cake [34].

2.3. Novel Assisting Cell Disruption Techniques

2.3.1. Microwave-Assisted Extraction (MAE)

Microwave-assisted extraction (MAE) is a novel cell disruption technique that uses electromagnetic waves of frequency in the range from 300 MHz to 300 GHz [23]. These waves are absorbed by the matrix and converted into thermal energy, which heats the moisture inside the cells. This generates a high pressure on the cells' walls, increasing their porosity and thus facilitating the extraction of their compounds [35]. The main MAE parameters required for optimization and scaling up the extraction process are the sample solubility, the solid-liquid ratio, the extraction process time and temperature, the microwave power, the system agitation, the dielectric constant, and the dissipation factor [36–38]. In comparison with conventional technologies, MAE presents some advantages, such as a higher reproducibility in a shorter period of time, and a lower solvent and energy consumption. These advantages make MAE a suitable technology for extracting different compounds, such as proteins, carbohydrates, and antioxidant polyphenols [39]. Studies on the MAE of proteins from various biological sources are presented in Table 1. This Table shows a comparison of the yield values, protein content, and other characteristics of the resulting products with those obtained from traditional extraction technologies, such as solvent extraction or steam infusion, among others. According to these studies, the use of MAE results in higher plant protein yields compared to the standard alkaline procedure [26,40]. Electromagnetic microwaves also provide other potential advantages over conventional hydro-thermal treatments, such as uniform heating, an enhanced extraction rate, lower solvent consumption, and a higher extraction speed [3,41]. Moreover, the improvements of functional properties and the digestibility of extracted proteins when using MAE have been reported [26,42]. Thus, MAE has been recommended as an approach to extracting protein from structurally rigid biological samples, which are difficult to digest using enzymes and/or ultrasound waves, such as bran, or other by-products of the milling industry (i.e., sesame, rice, and wheat) [43–45].

2.3.2. Ultrasound-Assisted Extraction (UAE)

Although ultrasound-assisted extraction (UAE) has been studied since the 1950s as a suitable approach to obtain proteins from natural sources, its application in food science is very recent, being recognized as a clean and novel technology. UAE is based on the propagation of pressure oscillations in a liquid medium at the speed of sound, which results in the formation, growth, and collapse of microbubbles, allowing cell disruption and a mass transfer to the medium [46]. The bubbles generated are relatively large and their collapse provokes cell wall breakup, a reduction of in particle size, and a mass transfer across cell membranes, allowing the extraction of substances from the medium. UAE's performance is affected by different parameters, such as the food matrix, extraction solvent, exposure time and temperature, ultrasound frequency, power, amplitude, and the type of equipment used [47]. As a green technology, UAE is energy efficient, easy to install, with minimal environmental impact, and its maintenance costs are low. Moreover, it requires a low investment and shorter extraction times, thus reducing the process time and the associated costs [48].

Thus, as a fast, cost-effective, and environmentally friendly technology, UAE has been used to extract and modify vegetable proteins, improving the efficiency of the extraction process. Recently, the existing evidence for the UAE of plant-based protein has been summarized by Rahman et al. [47]. Moreover, comparison between UAE and conventional technologies used to extract proteins from vegetal and animal food sources is shown in Table 1. Ochoa-Rivas et al. compared alkali extraction, MAE, and UAE to extract protein from peanut flour. The highest protein yield resulted from UAE, which also improved the techno-functional properties (the water absorption, foaming and emulsifying activities) and the in vitro protein digestibility of the extracted proteins [26]. Furthermore, UAE was recently applied to extract individual arachin and conarachin from defatted peanut protein, resulting in an increase in the extraction yield, a shortening of the extraction time and temperature, and an improvement of the emulsifying properties of arachin [49]. Similarly, increases in the protein yield and the enhancement of the functional and biological properties were reported after the UAE of proteins from pea and brewer's spent grain proteins [50,51]. Although the number of studies applying UAE to extract proteins from animal sources is more limited, this technology has been also recognized as a means to increase the extraction yield and improve the functional properties of extracted proteins from chicken liver and common carp (*Cyprinus carpio*) byproducts [52,53].

Despite being considered a novel and suitable approach to extracting proteins from different food materials, UAE has been associated with some weaknesses, such as the formation of radicals responsible for the release of degradation products affecting protein quality, resulting in protein oxidation, loss of aroma, changes in protein color, structure changes, texture alterations, free radical formation, and a metallic flavor [54].

Table 1. Comparison between conventional and emergent assisting cell disruption technologies (microwave and ultrasound-assisted) for the extraction of food proteins.

Food Protein Source	Conventional Extraction	Microwave-Assisted	Ultrasound-Assisted	Reference
		Results of the Extraction Process		
Rice bran	Extraction yield: 12.85% Protein content: 75.32% Extraction time: 60 min	Extraction yield: 15.68% Protein content: 79.98% Extraction time: 2 min		[41]
	Protein yield: 2.92%	Protein yield: 4.37% Protein content: 71.27%		[40]
Rice	Extraction yield: 38.0% Protein purity: 64.12%		Extraction yield: 65.0–86.0% (combined with α-amylase degradation) Protein purity: 77.47–92.99% Higher solubility, emulsifying activity and foaming capacity	[55]
Sesame bran	Protein content: 24.5% TPC: 3.45 mg GAE/g	Protein content: 43.8 to 61.6% (91.7% by MAEE) TPC: 4.20 mg GAE/g (8.04 mg GAE/g by MAEE) Highest recovery of antioxidant compounds		[44]
	Protein yield: 24.5% (alkaline extraction)-79.3% (enzymatic-assisted extraction)		Protein yield: 59.8 (ultrasound-assisted)-87.9% (combined with enzymatic treatment)	[45]

Table 1. *Cont.*

Food Protein Source	Results of the Extraction Process			Reference
	Conventional Extraction	**Microwave-Assisted**	**Ultrasound-Assisted**	
			Protein yield: 58.0% (vacuum-ultrasound assisted)-65.9% (vacuum-ultrasound assisted enzymatic extraction) Higher total phenolic capacity and antioxidant capacity	[56]
Peanut flour	Protein yield: 42.4%	Protein yield: 55.0% Improvement of water absorption, foam activity, emulsifying activity, and in vitro digestibility	Protein yield: 57.6% Improvement of water absorption, foam activity, emulsifying activity, and in vitro digestibility	[26]
Defatted peanut protein			Arachin extraction yield: 37.53% Conarachin extraction yield: 7.57% Shortening of the extraction time and temperature Improvement of emulsifying properties of arachin	[49]
Defatted wheat germ protein	Extraction yield: 24.0–37.0%	Extraction yield: 45.6% (combined with reverse micelles)		[43]
Pea protein	Extraction yield: 71.6%		Extraction yield: 82.6% Shortening of extraction times and reduction of water consumption Improvement of functional properties and biological activities	[50]
Alfalfa protein			Extraction yield: 14.5% (Ultrasound-ultrafiltration-assisted alkaline ioelectric precipitation) Potein content: 91.1 g/100 g Increase of solubility, water-holding and oil-binding capacities Reduction of emulsifying and foaming properties	[51]
Brewer's spent grain protein	Extraction yield: 45.71%		Extraction yield: 86.16% Protein purity: 57.84% Enhancement of the fat absorption capacity, emulsifying and foaming properties	[57]
Soy milk	Extraction yield: 3.86% (steam infusion) Protein content: 7.38%	Extraction yield: 4.83% Protein content 13.12% Improvement of characteristics of soy milk Increase of protein solubility and digestibility		[58]
Soy okra	Extraction yield: 0.35% (steam infusion) Protein content: 25.0%	Extraction yield: 0.23% Protein content 18.5%		[58]
Jackfruit leaves	Protein content: 8.41%	Protein content: 9.56%	Protein content: 9.63%	[59]

Table 1. *Cont.*

Food Protein Source	Results of the Extraction Process			Reference
	Conventional Extraction	Microwave-Assisted	Ultrasound-Assisted	
Eurycoma longifolia roots	Extraction yield: 9.76% in 38 min (heat assisted)		Extraction yield: 9.54% in 5 min	[60]
Coffee silverskin	Protein yield: 24.35% (alkali extraction)-32.52% (sequential alkaline-acid extraction)	Protein yield: 43.53%	Protein yield: 14.04%	[61]
Dolichos lablab L. protein	Extraction yield: 40.95%		Extraction yield: 69.98% Enhancement of functional characteristics and antioxidant capacity of the protein	[62]
Common carp by-products		Extraction yield: 0.82–1.27% Protein content: 87.63 to 88.19% Reverse correlation between the extraction time and the gel strength and viscosity of gelatin	Extraction yield: 19.80–27.0% Protein content: 86.15 to 90.21% Decrease of the gel strength and viscosity of gelatin	[53]
Duck feet gelatin	Extraction yield: 51.83% (water bath) and 22.06% (electric pressure cooker)	Extraction yield: 17.58% Improvement of gel strength, melting point, and viscosity		[42]
Bighead carp	Protein yield: 19.15–36.39% (water bath) Protein content: 84.15–88.67%	Protein yield: 30.94–46.67% Protein content: 89.17–91.85%		[63]
Chicken liver protein	Extraction yield: 43.5% Protein content: 63.9%		Extraction yield: 67.6% Protein content: 61.8% Improvement of the water/oil holding capacity and emulsifying properties	[52]

TPC: total phenolic content; GAE: gallic acid equivalent; MAEE: Microwave-assisted enzymatic extraction.

In addition, when UAE is applied at high intensities, it generates heat which provokes protein denaturation and unfolding. Thus, the intensity and synergy of ultrasound need to be optimized before their application. Recently, UAE has been combined with other technologies to reduce its limitations and improve the extraction of food proteins. Among these combinations, those including UAE and EAE have been reported to be effective to extract target compounds with the advantages of enhanced extraction yield, reduced extraction time, and physiological activity. Görgüç et al. compared alkali-EAE with ultrasound-EAE, reporting an increase in the sesame bran protein yield from 79.3% to 87.9% [45]. These authors also found an increase in the protein yield, phenolic content, and antioxidant capacity of the protein extracts when vacuum-ultrasound-EAE was applied to the sesame bran [56].

2.3.3. Pulsed Electric Field and High Voltage Electrical Discharge Extraction

Pulsed electric field (PEF) and high voltage electrical discharge (HVED) are emerging non-thermal technologies that use high voltage to generate an electric field to perform an extraction. Both technologies are based on the pores formed on the cell membrane (electroporation) resulting from the application of an electric field, thus allowing the extraction of intracellular components through a diffusion process [64]. In PEF technology, a material is placed between two electrodes through which high voltage pulses (from 100 to 300 V/cm to 20–80 kV/cm) are applied in short time periods. Under the effect of PEF, the cell membrane is electrically pierced, losing its permeability in a reversible or irreversible manner depending on the electrical parameters, the cell characteristics (size, age, and shape), and the pulsing media composition [65]. These features have made PEF into an extensive technology for microorganisms' inactivation, recovering high-value compounds, like proteins and polyphenols, and improving freezing and drying processes [66]. In HVED technology, a current of high-voltage electrical discharge is applied between two electrodes forming a plasma channel where energy is then presented directly in a liquid. Because PEF and HVED are non-thermal techniques, they present some advantages over conventional heat-assisted extraction techniques, such as their ability to preserve thermolabile food constituents as proteins and increasing the quality of the extracts during processing and throughout the storage period [13]. Moreover, the low energy consumption of these extraction processes agrees with the principles of green extraction [67].

Recent studies have applied PEF and HVED to recover proteins from different natural sources. For example, Roselló-Soto et al. compared the effects of PEF, HVED, and ultrasound pre-treatments before the extraction of intracellular compounds from olive kernels [68]. In this study, HVED technology was found to be more effective than others in terms of the energy and treatment time required to extract the phenolic compounds and proteins. Sarkis et al. also reported on the efficiency of the PEF and HVED approaches in increasing the yield of polyphenol, lignin, and protein extracts from sesame cake, and in improving the kinetics of diffusion [69]. More recently, PEF technology was applied to release and extract proteins from the microalgae *Chlorella vulgaris* and *Neochloris oleoabundans*. The protein yield was much lower than that obtained using mechanical approaches with a higher energy input [70]. Although PEF technology has been associated with changes in the solubility and functional properties of extracted proteins, these modifications can be beneficial. Thus, Zhang et al. reported that the PEF treatment of canola (*Brassica napus*) seeds increases the solubility, emulsifying, and foaming properties of the extracted proteins [71]. Changes in the secondary structure and in the antioxidant activity of peptides released from pine nut proteins have also been reported [72].

2.3.4. High Hydrostatic Pressure-Assisted Extraction

High hydrostatic pressure (HHP) technology is based on the application of an isostatic pressure ranging from 100 to 1000 MPa transmitted instantaneously and uniformly through a fluid, generally water. This pressure provokes the deformation of cells and the damage of their membranes and protein structures, thus allowing for the penetration of the solvent within cells and increasing the transfer rate of its intracellular components [73]. HHP processing, also known as "cold pasteurization", has been traditionally used in the food industry to reduce the microbial charge and improve the shelf-life of different food systems [74]. However, in recent years, HHP processing has been extended to other innovative uses, such as the selective recovery of phenolic compounds, polysaccharides, fats, and proteins, among others [75–77]. Moreover, HHP processing has become one of the most efficient methods for modifying the properties of proteins, such as thermal and rheological properties, gelation solubility, water holding and foaming capacity, stability, surface hydrophobicity, emulsifying activity, and stability [78]. It also induces the reversible denaturation of native proteins and the modulation of protein-protein and protein-solvent interactions, stimulating the formation of oligomeric and aggregated species that can negatively affect the digestibility of the protein [79,80]. However, in some cases, the alteration of the food proteins profile caused by HHP results in the higher exposure of susceptible digestion sites and, consequently, in a higher hydrolysis and yield of hydrolytic products, thereby reducing both time and costs [81]. In addition to its ability to modify proteins, HHP also shows an ability to modulate the conformation and activity of different enzymes as well as their interaction with their target proteins [82]. Additionally, in the recent years HHP has been employed in combination with enzymatic treatments to hydrolyze proteins from different food sources, such as kidney and pinto bean (*Phaseolus vulgaris*), soy and flaxseed (*Linum usitatissimum*) protein, and whey protein isolate, and to release bioactive peptides (see the review of Ulug et al. [83]). Since the efficiency of hydrolysis depends on the protein system, pressurization conditions, and the enzyme used, optimizing these parameters to produce the maximum amount of bioactive peptides from a complex food matrix is essential to accelerate the development of bioactive peptides-based food products.

3. Release of Bioactive Peptides

Bioactive peptides are sequences of amino acids inactive within the source protein. Once released, they can exert antioxidant, anti-inflammatory, antihypertensive, anti-obesity, antimicrobial, and immunomodulatory biological activities. These activities suggest their potential as novel, safe, and effective ingredients for functional foods and/or nutraceuticals to prevent and manage chronic disorders. Enzymatic hydrolysis and fermentation are the most known conventional methods to release bioactive peptides from their source protein. However, their use in the food industry is limited due to some disadvantages. Acid and alkaline hydrolysis are low-cost methodologies, but they affect some amino acid residues, thus resulting in the loss of nutritional and biological value of the released peptides [8]. Enzymatic proteolysis is a better alternative for hydrolyzing proteins due to the enzyme's specificity, but it can be expensive, time consuming, and require the use of acids and bases for pH control [84]. In the case of fermentation, it offers the advantage of removing hyper-allergic or antinutritional factors, but the costs are also relatively high [8]. Moreover, to produce protein hydrolyzates and bioactive peptides, efficient and scalable methods should be used to avoid the use of harmful chemicals and solvents, costly enzymes, and long processing times [21]. Therefore, in recent years, novel technologies such as ultrasounds, microwave-assisted processing, HHP, PEF, and SW hydrolysis are being explored to enhance the production of bioactive peptides [83] (Figure 1).

In addition to extracting proteins, microwave radiation is capable of improving proteolysis and releasing low molecular weight peptides when combined with enzymatic hydrolysis, thereby increasing the bioactivities of the resultant hydrolyzates. Thus, in comparison to thermal processing, MA-hydrolysis has significantly increased the dipeptidyl peptidase IV (DPP-IV) and ACE inhibitory activities of cricket (*Gryllodes sigillatus*) protein hydrolyzates [85], and the antioxidant capacity of chia (*Salvia hispanica*) protein hydrolyzates [86]. Similarly, this combination has been used to prepare bioactive peptides from bovine serum albumin, ginkgo (*Ginkgo biloba*) nuts [87], milk protein concentrate [88], bighead carp (*Aristichthys nobilis*) [89], fish protein [90], trout frame (*Oncorhynchus mykiss*) [91], and collagen sea cucumber (*Acaudina molpadioides*) [92]. Microwave radiation facilitates the exposure of cleavage sites of the protein to the enzyme action; thus, the hydrolysis efficiency, processing time, and reproducibility are generally improved by this technology. Other advantages of microwave processing over conventional methods include its simple handling and low cost [93].

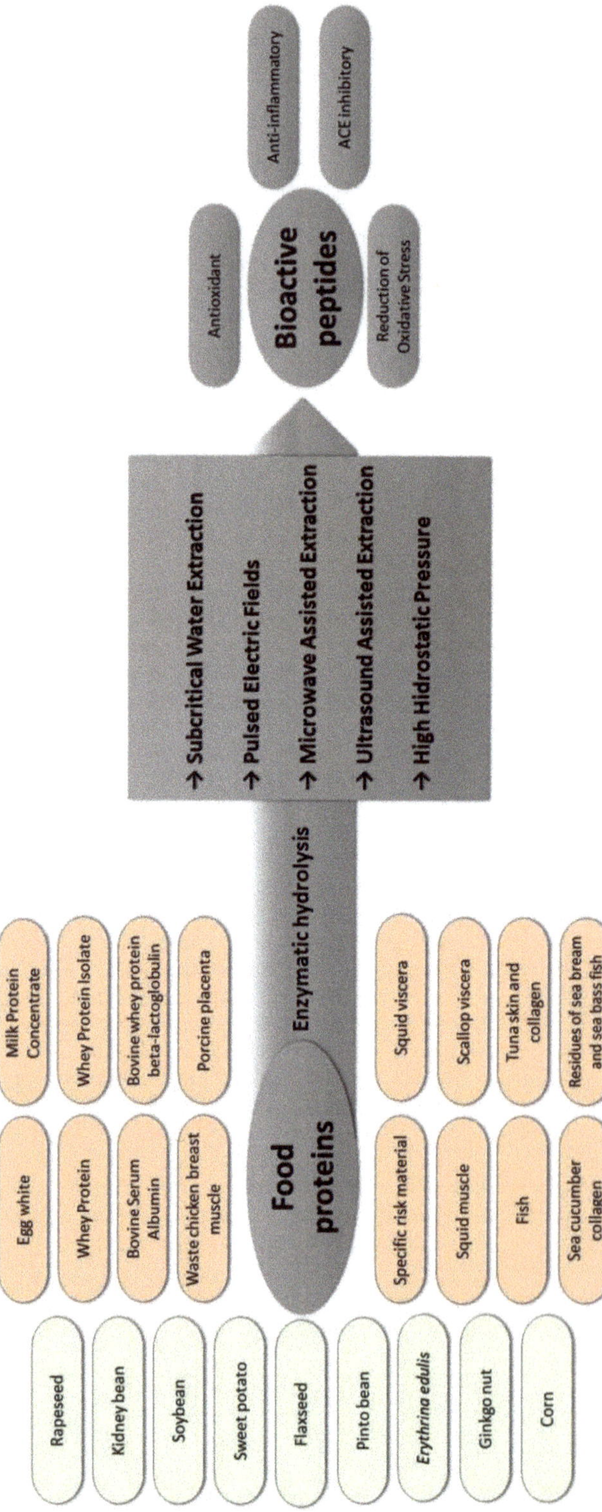

Figure 1. Hydrolysis of food sources assisted by novel and green technologies to release bioactive peptides.

Although the application of ultrasounds alone is not enough to break peptidic bonds, this technology is capable of increasing the ability for enzymes into enter peptide bonds; thus, its combination with enzymatic hydrolysis is recognized as a suitable means to improve the production of bioactive peptides. This combination also shows other advantages over conventional hydrolysis, such as a speedier energy and mass transfer, diminished temperature and time, a higher process control, and a higher selectivity of the extraction [39]. Several food proteins have been used as a source of bioactive peptides following the application of UA-hydrolysis. Thus, more potent ACE inhibitory [94] and antioxidant [95] peptides were released after applying an ultrasound step before hydrolysis of wheat protein. More recently, the improvement of the ACE inhibitory activity of rapeseed protein hydrolyzates has been demonstrated [96]. Ultrasound has also been employed to increase the bioactivities of hydrolyzates from animal proteins such as milk and whey protein [88,97] and ovotransferrin [98].

HHP is another green technology employed to increase the production process of bioactive peptides due to its ability to denature protein and improve the accessibility of enzymes within susceptible cleavage sites [83]. Within the animal kingdom, HHP-assisted enzymatic hydrolysis has been applied to whey protein isolate [99] and beta-lactoglobulin [100] to release bioactive peptides. Similarly, this technology has been employed to produce bioactive hydrolyzates from pinto beans [101], kidney beans [102], soy protein [103], and flaxseed protein [104].

Because of its ability to break non-covalent bonds, such as hydrogen bonds, hydrophobic interactions, and covalent bonds, PEF is considered as an environmentally friendly choice for producing antioxidant extracts from food byproducts, such as fish heads, bones, and gills [105]. PEF has also been recognized as useful for hydrolyzing food proteins, such as egg white proteins [106–108] and soybean proteins [109], resulting in the release of small weight antioxidant peptides. Although this methodology shows some advantages over conventional hydrolysis, such as shorter treatment times and lesser energy consumption, its current use is still limited due to its' high investment costs [110].

SW-hydrolysis (SWH) is a clean and fast protein hydrolysis alternative to acid, alkali, and enzymatic hydrolysis methods. It implies the application of high pressure to maintain water in its liquid state at temperatures above its boiling point. The high temperature results in high diffusion, low viscosity, and low surface tension, allowing the hydrolysis of proteins in shorter times [111]. Thus, because of its non-toxic and non-flammable attributes, SWH has been used in recovering a large variety of high value-added bioactive protein hydrolyzates and peptides from different animal and vegetable sources (Table 2). Despite the fact that SWH can be applied as a pre-treatment process before enzymatic hydrolysis to improve the biological activities of food proteins [112], it can also be used without using enzymes, thus reducing the process time and costs and avoiding the use of hazardous solvents and chemicals [113]. However, some disadvantages, such as high infrastructure costs and the necessity to optimize multiple process conditions to increase the efficiency of SWH, limit the current application of this methodology [83].

4. Supercritical CO$_2$ in Protein Extraction and Processing

Supercritical carbon dioxide (SC-CO$_2$) has opened a wide range of new alternatives in food technology. Many studies and applications of SC-CO$_2$ in the processing of certain target food components, such as oils, fatty acids, phytosterols, alkaloids, carotenoids, and flavonoids, were widely reported in the last few decades. Proteins were not the exception, despite the complexity of their structure and properties. Several applications of SC-CO$_2$ in protein processing are related with the elimination of oils and other lipophilic substances from protein-based matrices, such as the recent works reported by different authors [114–116]. In these works, the supercritical step is just a pre-treatment and the SC-CO$_2$ is used as an extractive solvent to eliminate the oily substances prior to the protein isolation from the food matrix (canola seeds, quinoa seeds and corn germ, respectively).

Besides extraction, SC-CO$_2$ can be applied in other protein-based processes, such as the precipitation of protein particles, the separation of peptides, or the improvement of protein functionalities taking into advantage the structural and conformational modifications that can be attained by their exposure to high pressure CO$_2$.

Table 2. Subcritical water assisted hydrolysis of natural proteins to release bioactive hydrolyzates and peptides.

Natural Source	Hydrolysis Conditions	Results	Reference
Whey protein isolate (WPI)	WPI:water ratio = 60 g/L Different temperatures and times (experimental design)	Effective and fast (<60 min) WPI hydrolysis Generation to the highest total amino acid and lysine concentration at 300 °C for 40 min	[84]
Bovine seroalbumin	Seroalbumin:water ratio = 10 mg/mL Different temperatures and times	Greatest release of free amino groups and maximum amount of amino acids at 280 °C High generation of alanine and glycine	[117]
Hemoglobin, bovine seroalbumin, and β-casein	Protein:water ratio = 1 mg/mL Different temperatures and times	High protein sequence coverages (>80%) comparable to those obtained by trypsin digestion Favored cleavage of the Asp-X bond under mild conditions (160 °C) for three proteins Favored cleavage of the Glu-X bond under 160 °C and 207 °C (β-casein) and 207 °C (seroalbumin)	[118]
Scomber japonicus muscle protein	Different temperatures	Highest degree of hydrolysis at 140 °C for 5 min Highest antioxidant activity at 140 °C for 5 min and tyrosinase inhibitory activity at 200 °C for 15 min	[119]
Mackerel (*Scomber japonicus*)	Collagen:water ratio = 1:200 (*w:v*) Hydrolysis time = 3 min	Release of small and potent antioxidant peptides	[111]
Bigeye tuna skin	Bacterial collagenolytic protease-extracted collagen:solvent ratio = 1:200 (*w:v*) Hydrolysis time = 3 min	Efficient hydrolysis of collagen Release of small (<425 Da) and potent antioxidant and antimicrobial peptides	[120]
Tuna skin collagen and skin	Sample:liquid ratio = 1:200 (collagen) and 1:50 (skin) Different temperatures Hydrolysis time = 5 min	Highest antioxidant and antimicrobial activity at 280 °C Release of low molecular weight peptides (<600 Da) and/or free amino acids associated with the bioactivity	[121]
Atlantic cod (*Gadus morhua*) frames	Different temperatures Hydrolysis time = 30 min	Release of smaller peptides at high temperatures (250 °C) Potent anti-inflammatory potential of hydrolyzates in Caco-2 cells	[122]
Comb penshell (*Atrina pectinata*) viscera	Powder:water ratio = 30 g/L Different temperatures Hydrolysis time = 15 min	Release of small (<1000 Da) peptides at temperature higher than 200 °C Release of antioxidant and anti-hypertensive peptides	[123]
Green seaweed (*Ulva* sp.)	Algae:seawater ratio = 8% (*w:w*) Hydrolysis time = 40 min	Efficient protein extraction as starting material for fermentation with *E. coli* and *S. cerevisiae*	[124]
Algae Laver (*Pyropia yezoensis*)	Powder:water ratio = 1:20 (*w:v*) Different temperatures Hydrolysis time = 30 min	Extraction of the maximum amount of amino acids at 120 °C Extraction of potent antioxidant compounds	[125]
Soy protein	Powder:water ratio = 62.5 g/L Different temperatures	Highest amino group content and yield at 190 °C Influence of the temperature on color parameters Release of small peptides Inhibitory effects on murine macrophages viability	[126]

4.1. Precipitation of Micro and Nano Protein Particles

Several SC-CO$_2$ techniques are currently available for particle formation, encapsulation and the drying of a wide type of materials, indicating the flexibility of SC-CO$_2$ technology which offers worthy advantages, such as the control of the particle size, size distribution, and morphology, in comparison with conventional technologies (spry-drying, jet milling, freeze drying, and coacervation). Food lipids, carbohydrates, proteins, and minor components have been processed using these SC-CO$_2$ techniques with the major purpose to deliver bioactive components.

Table 3 shows several protein-based SC-CO$_2$ processes reported in the literature, categorized according to the overall objective of the supercritical approach, i.e., the formation of micro/nanoparticles and the use of SC-CO$_2$ for drying and atomization.

Table 3. Examples of protein-based ingredients obtained using pressurized fluid technologies.

Protein	SC-CO$_2$ Technique	Abbreviation	Reference
	Particle formation		
Lysozime	Precipitation with a Compressed fluid Antisolvent	PCA	[127]
Lysozime	Solution Enhanced Dispersion of Solids	SEDS	[128]
Whey protein isolate (WPI)	Gas Anti-Solvent	GAS	[129]
Bovine serum albumin (BSA)	Particles from Gas Saturated Solutions	PGSS	[130]
Zein	Supercritical Anti-Solvent	SAS	[131]
Lysozime	Supercritical Fluid Extraction of Emulsions	SFEE	[132]
SC-CO$_2$-assisted drying and atomization			
Gelatin	Particles from Gas Saturated Solutions for drying	PGSS drying	[133]
Lysozime + sugars	Particles from Gas Saturated Solutions for drying	PGSS drying	[134]
Lysozime	Expanded Liquid Anti-Solvent	ELAS	[135]
Bovine serum albumin (BSA)	Expanded Liquid Anti-Solvent	ELAS	[136]
Lysozime	Supercritical Assisted Atomization	SAA	[137]
Bovine serum albumin (BSA)	Supercritical Assisted Atomization with Hydrodynamic Cavitation Mixer	SAA-HCM	[138]
Lysozime	Supercritical Assisted Atomization with Hydrodynamic Cavitation Mixer	SAA-HCM	[139]
Trypsin and trypsin-chitosan	Supercritical Assisted Atomization with Hydrodynamic Cavitation Mixer	SAA-HCM	[140]
SC-CO$_2$-assisted impregnation			
Bovine haemoglobin (bHb)	Supercritical Solvent-Assisted Impregnation	SSI	[141]
Soy protein	Supercritical Solvent-Assisted Impregnation	SSI	[142]

SC-CO$_2$ techniques using solvents such as dimethyl sulfoxide (DMSO) contradicts the main advantage of this technology for food applications. For example, Moshashaée et al. determined 20 ppm of residual DMSO solvent in the Solution Enhanced Dispersion of Solids (SEDS) precipitation of lysozyme [128]. The challenge is circumventing the use of solvents or limiting their use to food-grade solvents. This is the case of the friendly whey protein fractionation process developed by Yver et al. to produce enriched fractions of α-lactalbumin (α-LA) and β-lactoglobulin (β-LG) from a commercial whey protein isolate (WPI) [129]. SC-CO$_2$ was injected into the vessel containing the WPI de-ionized water solution, and the pH modification produced due to the SC-CO$_2$ dissolution allowed the fractionation of the proteins. A solid α-LA-enriched phase was selectively precipitated at pH 4.4–5.0 and separated from the liquid β-LG-enriched fraction. After the supercritical treatment, these fractions were ready-to-use and did not contain salt, acid, or other chemical contaminants. The highest α-LA purity was 61% (α-LA initial concentration in WPI was 18%) with 80% of α-LA recovery in the solid fraction, and was obtained at 60 °C, 83 bar and 5% of proteins in WPI. Lima et al. described a continuous flow reactor for the protein fractionation process, testing pressures in the range of 80–240 bar and temperatures of 55–65 °C [143]. The initial ratio α-LA/β-LG in the WPI was 1:2.7 and the combination 80 bar and 55 °C was found to be the best condition to obtain α-LA (α-LA/ β-LG =3.84 and 20.9% precipitation yield). These process conditions also resulted in a β-LG-enriched fraction. Recently, the authors reported a techno-economic assessment of this continuous α-LA/β-LG supercritical fractionation [144].

Other remarkable reports concerning the protein particle formation using SC-CO$_2$ technique are those published by Perinelli et al., Zhong et al., and Kluge et al. [130–132]. Avoiding the use of an organic solvent, the encapsulation of bovine seroalbumin (BSA) using biodegradable copolymers has been carried out applying particles from gas-saturated solutions (PGSS) [130]. Zhong et al. developed a supercritical anti-solvent (SAS)-based process to manufacture, generally recognized as safe, (GRAS) delivery systems to release antimicrobials, thereby enhancing the shelf-lives of foods [131]. Corn zein was used as the carrier material and egg white lysozyme dissolved in 90% aqueous ethanol was microencapsulated. The authors revealed a long-time continuous release of antimicrobials of lysozyme at neutral pH conditions in the presence of salt. On the other hand, Kluge et al. studied the supercritical-fluid-extraction of emulsions (SFEE) process for the manufacturing of lysozyme—poly-lactic-co-glycolic acid (PLGA) composite particles for the delivery of lysozyme protein [132]. SFEE combines the efficiency of SC-CO$_2$ extraction with the facilities of water in oil (w/o) or water in oil in water ($w/o/w$) double emulsion methods. The SC-CO$_2$ and the emulsion feed streams are mixed at the inlet of the reactor in a two-coaxial nozzle. The particles are formed due to the organic solvent extraction from the emulsion droplets and remain suspended in the continuous water phase. Different encapsulation methods were tested and evaluated using ethyl acetate as organic solvent but the encapsulation efficiencies were lower than 50%. The authors concluded that despite the efficacy of SFEE process to produce solvent-free PLGA particles with small size and homogenous distribution, the encapsulation of drugs is more challenging for very hydrophilic compounds, such as peptides and proteins.

The use of different SC-CO$_2$ techniques for drying and atomization in protein-based processes was mainly limited to pharmaceutical applications, but research and development in the area of food-related products, including natural health ingredients, is increasing rapidly in recent years. Protein-based ingredients may be the single protein or multicomponent composite systems. The micronization of a single protein can improve bioavailability due to the increase of specific surface area, whereas composite systems refer to micro or nano particles where a certain active component is coated with the protein. In this respect, both proteins and polysaccharides are food-grade GRAS coatings and are preferred in food ingredient processing, in comparison with synthetic polymers (i.e., polyethylene glycol, polylactic acid, PLGA), which are especially selected as excipients for drug delivery in pharmaceuticals.

The supercritical drying of aqueous based solutions requires the use of very large quantities of CO$_2$ due to the low solubility of water in SC-CO$_2$ [145], or very high temperatures (>120 °C), as in the case of PGSS-drying [145] or PGX (gas-expanded liquids) [146].

Nuchuchua et al. have illustrated PGSS-drying as a scalable organic solvent free SC-CO$_2$ spray drying process for producing dry protein/sugar formulations (1:10 and 1:4 w/w ratios) [134]. Reibe et al. introduced the drying, micronization, and formulation of high molecular mass gelatine:aqueous gelatine solutions with a dry mass content of up to 50% w/w, which were pulverised and dried with minor hydrolysis degradation during PGSS-drying supercritical processing [133].

In PGX, liquid ethanol expanded by the dissolution of CO$_2$ is an appropriate choice for polysaccharides, or proteins drying. SC-CO$_2$ removes water but also can act as an antisolvent for the precipitation of the biopolymer. The process parameters must be selected to ensure a homogeneous single liquid phase of the ethanol-water-CO_2 ternary system. Thus, the liquid-liquid mixing operation (aqueous biopolymer solution with CO_2-expanded ethanol) avoids mass transfer problems involved in spraying processes. The precipitated polysaccharide, or protein, is collected on a filter after the liquid is drained by passing SC-CO_2. In this respect, expanded liquid anti-solvent (ELAS) protocol was used to denote the protein drying processes using CO_2-expanded ethanol, acetone, and isopropyl alcohol. De Marco et al. used ELAS with CO_2-expanded ethanol to yttrium acetate and BSA water-soluble materials, to produce micro and nanoparticles [136]. At the appropriate operating conditions, BSA spherical and non-coalescing particles were produced, with a narrow particle size distribution and mean diameter in the range 0.5–2.0 μm.

The supercritical assisted atomization (SAA) is another micronized technique that uses a thermostatic saturator to solubilize CO_2 in the drug solution, and a thin wall injector to induce the atomization of the solution into the precipitator vessel. The fast release of CO_2 from the primary formed droplets causes decompression and forms smaller secondary droplets. The SAA process was successfully applied to micronize antibiotics and polymers from either organic solvents or water, including protein micronization [137,147]. A hydrodynamic cavitation mixer (HCM) added to the saturator can intensify mass transfer between CO_2 and the liquid solution. The SAA-HCM process was used [139] to produce lysozyme microparticles with a controlled particle size distribution. The particles were well defined (no agglomerates), spherical, and with 0.2–5.0 μm diameters at the optimum operating conditions. Bioactivity assays showed the maintenance of 85% of lysozyme original activity. Furthermore, Wang et al. used this technique to prepare BSA microparticles from water solutions, obtaining different morphologies with particle diameters in the range of 0.3–5.0 μm [138]. Moreover, Shen et al. applied SAA-HCM to prepare micrometric particles of trypsin from aqueous solutions, obtaining several morphologies depending on the process conditions, while also analyzing the structural stability of the protein to conclude that trypsin retained above 70% of its biological activity [140]. Moreover, polymer chitosan was used to prepare trypsin composite microparticles, obtaining spherical microparticles with a homogeneous size distribution with 90% efficiency.

Supercritical solvent-assisted impregnation (SSI) is a more recent protein-based application of SC-CO_2 technology. Trivedi et al. investigated the coating of protein-immobilised particles with myristic acid using SC-CO_2 processing at low temperatures, in order to prevent thermal degradation of the protein (bovine haemoglobin, bHb) [141]. A solid core drug delivery system was prepared using bHb immobilisation on mesoporous silica followed by supercritical myristic acid coating at 43 °C and 100 bar. Protein particles were also coated via solvent evaporation to compare the protein release. In both methods, myristic acid coating provided good protection in gastric fluid media and limited the bHb release for the first two hours. After the change to intestinal fluid media, the protein release reached 70% within three hours. The release from supercritical samples was slower than with solvent evaporation formulations, indicating a superior myristic acid coverage, in addition to the protein conformation remaining unchanged after the release. Similarly, soy protein microparticles were coated with chia oil [142] using SC-CO_2. A good encapsulation efficiency was attained in the range of process conditions 100–160 bar, 40–60 °C and 0.0–0.1 *w/w* for the ethanol:oil ratio (ethanol was used as a co-solvent to increase the oil solubility in the SC-CO_2). The chia oil-loaded microparticles showed a spherical shape, no pores or fissures, sizes between 1 and 10 μm, and a homogeneous oil distribution. Furthermore, the hydroperoxide values and fatty acid profile indicated that the SSI process did not affect the chemical quality of the oil; the product showed an excellent oxidative stability, and almost all of the oil contained in the protein microparticles was released under gastro-intestinal conditions, remaining available for absorption.

It can be concluded that protein-based ingredients produced using SC-CO_2 technologies show good potential for further development, considering the growing demand for natural health products. Encapsulation of bioactive proteins is commonly accomplished using polysaccharides leading to dry powders which can be uniformly incorporated into different formulations even at low concentrations. In this way, efficient delivery systems can be designed. Varying supercritical processes parameters (temperature, pressure, flow rate ratios, nozzle diameter and depressurization rate, among others) the particle size, and morphology can be effectively controlled. Furthermore, low temperatures favour the handling of heat-sensitive substances, and circumventing the use of organic solvents, or the easy removal when they are used, is another key advantage of SC-CO_2 technology applied to the formation of bioactive protein-based microparticles.

4.2. Structural and Conformational Modifications Resulting from SC-CO_2 Treatments

Proteins and their products are used as ingredients in several food applications (dairy and bakery products, infant foods, and beverages) with the objective of enhancing nutritional value, creating emulsification, foaming, and antioxidant barriers, among others [148]. Since the functionality of food proteins are related to their structures, the structural modification of proteins can lead to new or improved functionalities [149]. In this respect, SC-CO_2 treatment was described as a green method to achieve protein modification [150]. SC-CO_2 offers many advantages in comparison with thermal processing methods, such as minimizing the alterations and quality of food [151]. Several studies have been reported, using SC-CO_2 for the inactivation of microorganisms and enzymes in liquid foods [152]. Despite many reports concerning the physical modification of polysaccharides under SC-CO_2 processing [153], fewer studies can be found concerning the effects of SC-CO_2 treatment on protein quality and functionality.

For example, Zhong et al. analysed the effect of SC-CO_2 treatment on the rheological properties of whey protein [154]. The authors concluded that temperature, pressure, holding time, protein concentration, and pH were the main operating parameters influencing the conformational, structural, and functionalities of whey protein isolate (WPI). Later, Xu et al. conducted a complete analysis of the physical, conformational, and structural properties of WPI treated with SC-CO_2 in comparison with the thermal treatment of WPI solution [150]. Increased turbidity of WPI treated with SC-CO_2 (50–60 °C at 20 MPa for 1 h) suggested more intensive WPI denaturation in comparison with thermal processing. Furthermore, dynamic light scattering measurements showed higher aggregates and a maximum mean particle size when WPI was processed with SC-CO_2 at 60 °C, indicating the partial aggregation of the dimer into the polymers. An analysis of the fluorescence emission spectra of proteins suggested changes in the polarity of the protein residue's microenvironment from a less polar to a more polar. The authors evidenced secondary and tertiary protein structure changes induced by SC-CO_2 treatment, resulting in physicochemical and functional properties of proteins.

The mechanisms of protein denaturalization due to SC-CO_2 treatment is actually a major topic of research due to its importance in cold sterilization methods. The complexity of the process has limited the possibility of attaining microstructural information related with microbial and/or enzyme inactivation by SC-CO_2. In a recent work, Monhemi and Dolatabadi used molecular dynamics simulation of SC-CO_2 pasteurization to conclude that the denaturation of proteins (myoglobin and lysozyme as models) is produced at the CO_2/water interface [155]. The protein migrates from a pure aqueous phase to the CO_2/water interface, releasing the hydrophobic cores to the CO_2 phase and the hydrophilic residues to the aqueous phase and then the protein is denatured to a flat and extended conformation. Several other works have been recently reported, aiming to explain the molecular mechanisms of protein denaturalization in SC-CO_2 [156,157], or describing novel food applications of SC-CO_2 enzyme inactivation. For example, Podrepšek et al. applied an SC-CO_2 treatment to inactivate the polyphenol oxidase enzyme in order to extend shelf life of a coarse-ground flour from whole wheat (graham flour) while preserving, at the same time, its high quality [158]. The effect of pressure on protein concentration and enzyme activity was significant, evidencing a 35% decrease in polyphenol oxidase enzyme activity after SC-CO_2 exposure at 300 bar for 24 h. Furthermore, the quality of the flour remained almost the same or even improved, indicating that the SC-CO_2-treated graham flour is still suitable for use in the bakery industry. Also, SC-CO_2 + ethanol extraction of yellow pea was found to enhance organoleptic attributes of the edible seed, reducing most chemical constituents, and lowering certain functionality qualities, such as pasting viscosities. Scanning electron micrographs indicated that protein-rich particles were adhered to the surface of starch granules of pea flour extracted with SC-CO_2, and this change may be responsible for the different functionality observed [157].

Thermal extrusion has been effectively used to improve the techno-functional properties of food proteins. The combination of temperature and shear weakens the structure of proteins and the extrusion process helps to partially unfold and denature proteins to modify their functionalities. SC-CO_2 extrusion is an effective technique to produce conformational changes in proteins, providing several advantages in comparison to steam extrusion cooking, such as low-temperature processing, formation of a uniform and porous product, and pH control. For instance, the use of milk protein concentrate (MPC) in extruded products is still a challenge because of its sensitivity to high temperature and shear. For example, Yoon et al. have demonstrated the improved physicochemical properties of MPC extruded with SC-CO_2 in comparison with steam extrusion [158]. Also, Gopirajah et al. applied SC-CO_2 extrusion to improve emulsifying properties of MPC, obtaining a better emulsion activity index, reduced creaming index, and similar viscosity as compared to commercial sodium caseinate [159].

Ding et al. showed that SC-CO_2 treatment can be a novel method to improve the foam ability of egg white protein [160]. At 9 MPa SC-CO_2 treatments, the protein foaming capacity increased 3.6 times in comparison with the original egg white. The protein electrostatic force was reduced due to the pH change and their particle size decreased. Accordingly, these effects resulted in an increase of protein hydrophobicity and viscosity.

5. Conclusions

Protein extraction is commonly considered to be a crucial step in contributing to the final greenness of the complete protein analysis process. In recent years, important advances have been achieved with the design, development, and application of novel methodological approaches for extracting high quality proteins at high yields. These new techniques also reduce the use of hazardous chemicals and solvents, and decrease the temperature and time conditions of the extraction process. Emerging eco-innovative extraction technologies are becoming a promising alternative to conventional methods for producing safe and nutritive proteins with preserved techno-functional properties. Furthermore, the combination of some of these methodologies with enzymatic hydrolysis has been demonstrated to improve the yield and biological properties of protein-derived peptides with the capability to be incorporated into functional foods and nutraceuticals. As presented in this review, although many advantages have been associated with the use of these technologies at a laboratory scale, the research is still in its infancy, thus further studies demonstrating their economical and environment suitability at an industrial scale are required.

Author Contributions: Conceptualization, T.F. and B.H.-L.; writing—original draft preparation, G.F.-O., T.F. and B.H.-L.; writing—review and editing, T.F. and B.H.-L. funding acquisition, B.H.-L. All authors have read and agreed to the published version of the manuscript.

Funding: This research was funded by the Spanish Ministry of Science and Innovation, projects PID2019-104218RB-I00/AEI/2015-66886R and PID2019-110183RB-C22/AEI/10.13039/501100011033.

Institutional Review Board Statement: Not applicable.

Informed Consent Statement: Not applicable.

References

1. Chakrabarti, S.; Guha, S.; Majumder, K. Food-derived bioactive peptides in human health: Challenges and opportunities. *Nutrients* **2018**, *10*, 1738. [CrossRef] [PubMed]
2. Kumar, M.; Tomar, M.; Potkule, J.; Verma, R.; Punia, S.; Mahapatra, A.; Belwal, T.; Dahuja, A.; Joshi, S.; Berwal, M.K.; et al. Advances in the plant protein extraction: Mechanism and recommendations. *Food Hydrocoll.* **2021**, *115*, 106595. [CrossRef]
3. Lee, S.Y.; Show, P.L.; Ling, T.C.; Chang, J.S. Single-step disruption and protein recovery from *Chlorella vulgaris* using ultrasonication and ionic liquid buffer aqueous solutions as extractive solvents. *Biochem. Eng. J.* **2017**, *124*, 26–35. [CrossRef]
4. Chen, R.; Wang, X.J.; Zhang, Y.Y.; Xing, Y.; Yang, L.; Ni, H.; Li, H.-H. Simultaneous extraction and separation of oil, proteins, and glucosinolates from *Moringa oleifera* seeds. *Food Chem.* **2019**, *300*, 125162. [CrossRef] [PubMed]
5. Du, M.; Xie, J.; Gong, B.; Xu, X.; Tang, W.; Li, X.; Li, C.; Xie, M. Extraction physicochemical characteristics and functional properties of Mung bean protein. *Food Hydrocoll.* **2018**, *76*, 131–140. [CrossRef]
6. Feyzi, S.; Varidi, M.; Zare, F.; Varidi, M.J. Fenugreek (*Trigonella foenum graecum*) seed protein isolate: Extraction optimization, amino acid composition, thermo and functional properties. *J. Sci. Food Agric.* **2015**, *95*, 3165–3176. [CrossRef]
7. Bernardi, S.; Corso, M.; Baraldi, I.; Colla, E.; Canan, C. Obtaining concentrated rice bran protein by alkaline extraction and stirring–Optimization of conditions. *Int. Food Res. J.* **2018**, *25*, 1133–1139.
8. Hou, F.; Ding, W.; Qu, W.; Oladejo, A.O.; Xiong, F.; Zhang, W.; He, R.; Ma, H. Alkali solution extraction of rice residue protein isolates: Influence of alkali concentration on protein functional, structural properties and lysinoalanine formation. *Food Chem.* **2017**, *218*, 207–215. [CrossRef]
9. Shao, D.; Atungulu, G.G.; Pan, Z.; Yue, T.; Zhang, A.; Fan, Z. Characteristics of isolation and functionality of protein from tomato pomace produced with different industrial processing methods. *Food Bioproc. Tech.* **2014**, *7*, 532–541. [CrossRef]
10. Mechmeche, M.; Kachouri, F.; Chouabi, M.; Ksontini, H.; Setti, K.; Hamdi, M. Optimization of extraction parameters of protein isolate from tomato seed using response surface methodology. *Food Anal. Methods* **2017**, *10*, 809–819. [CrossRef]
11. Mori, K.; Goto-Yamamoto, N.; Kitayama, M.; Hashizume, K. Loss of anthocyanins in red-wine grape under high temperature. *J. Exp. Bot.* **2007**, *58*, 1935–1945. [CrossRef] [PubMed]
12. Pickardt, C.; Eisner, P.; Kammerer, D.R.; Carle, R. Pilot plant preparation of light-coloured protein isolates from de-oiled sunflower (*Helianthus annuus* L.) press cake by mild-acidic protein extraction and polyphenol adsorption. *Food Hydrocol.* **2015**, *44*, 208–219. [CrossRef]
13. Cui, Q.; Ni, X.; Zeng, L.; Tu, Z.; Li, J.; Sun, K.; Chen, X.; Li, X. Optimization of protein extraction and decoloration conditions for tea residues. *Hortic. Plant J.* **2017**, *3*, 172–176. [CrossRef]
14. Abelson, J.; Simon, M. *Aqueous Two-Phase Systems*; Methods in Enzymology; Academic Press: New York, NY, USA, 1994; Volume 228.
15. Varadavenkatesan, T.; Pai, S.; Vinayagam, R.; Pugazhendhi, A.; Selvaraj, R. Recovery of value-added products from waste water using Aqueous Two-Phase Systems–A review. *Sci. Total Env.* **2021**, *778*, 146293. [CrossRef] [PubMed]
16. Saravana Pandian, P.; Sindhanai Selvan, S.; Subathira, A.; Saravanan, S. Optimization of aqueous two phase extraction of proteins from *Litopenaeus vannamei* waste by response surface methodology coupled multi-objective genetic algorithm. *Chem. Prod. Process Model.* **2019**, *15*, 20190034. [CrossRef]
17. Menegotto, A.L.L.; Fernandes, I.A.; Bucior, D.; Balestieri, B.P.; Colla, L.M.; Abirached, C.; Franceschi, E.; Steffens, J.; Valduga, E. Purification of protein from *Arthrospira platensis* using aqueous two-phase system associate with membrane separation process and evaluation of functional properties. *J. Appl. Phycol.* **2021**. [CrossRef]
18. Zhang, J.; Wen, C.; Zhang, H.; Duan, Y.; Ma, H. Recent advances in the extraction of bioactive compounds with subcriticalwater: A review. *Trends Food Sci. Technol.* **2020**, *95*, 183–195. [CrossRef]
19. Mlyuka, E.; Mbifile, M.; Zhang, S.; Zheng, Z.; Chen, J. Strategic applications and the challenges of subcritical water extraction technology in food industries. *Chiang Mai J. Sci.* **2018**, *45*, 1015–1029. Available online: https://www.thaiscience.info/Journals/Article/CMJS/10989357.pdf (accessed on 6 September 2021).
20. Marcet, I.; Álvarez, C.; Paredes, B.; Díaz, M. The use of subcritical water hydrolysis for the recovery of peptides and free amino acids from food processing wastes. Review of sources and main parameters. *Waste Manag.* **2016**, *49*, 364–371. [CrossRef] [PubMed]
21. Lamp, A.; Kaltschmitt, M.; Lüdtke, O. Protein recovery from bioethanol stillage by liquid hot water treatment. *J. Supercrit. Fluids* **2020**, *155*, 104624. [CrossRef]
22. Álvarez-Viñas, M.; Rodríguez-Seoane, P.; Flórez-Fernández, N.; Torres, M.D.; Díaz-Reinoso, B.; Moure, A.; Domínguez, H. Subcritical water for the extraction and hydrolysis of protein and other fractions in biorefineries from agro-food wastes and algae: A review. *Food Bioproc. Tech.* **2021**, *14*, 373–387. [CrossRef]
23. Pojić, M.; Mišan, A.; Tiwari, B. Eco-innovative technologies for extraction of proteins for human consumption from renewable protein sources of plant origin. *Trends Food Sci. Technol.* **2018**, *75*, 93–104. [CrossRef]
24. Sari, Y.W.; Bruins, M.E.; Sanders, J.P.M. Enzyme assisted protein extraction from rapeseed, soybean, and microalgae meals. *Ind. Crops Prod.* **2013**, *43*, 78–83. [CrossRef]
25. Liu, J.-J.; Gasmalla, M.A.A.; Li, P.; Yang, R. Enzyme-assisted extraction processing from oilseeds: Principle, processing and application. *Innov. Food Sci. Emerg. Technol.* **2016**, *35*, 184–193. [CrossRef]

26. Ochoa-Rivas, A.; Nava-Valdez, Y.; Serna-Saldívar, S.O.; Chuck-Hernández, C. Microwave and ultrasound to enhance protein extraction from peanut flour under alkaline conditions: Effects in yield and functional properties of protein isolates. *Food Bioproc. Technol.* **2017**, *10*, 543–555. [CrossRef]
27. Rommi, K.; Hakala, T.K.; Holopainen, U.; Nordlund, E.; Poutanen, K.; Lantto, R. Effect of enzyme-aided cell wall disintegration on protein extractability from intact and dehulled rapeseed (*Brassica rapa* L. and *Brassica napus* L.) press cakes. *J. Agric. Food Chem.* **2014**, *62*, 7989–7997. [CrossRef] [PubMed]
28. Rommi, K.; Ercili-Cura, D.; Hakala, T.K.; Nordlund, E.; Poutanen, K.; Lantto, R. Impact of total solid content and extraction pH on enzyme-aided recovery of protein from defatted rapeseed (*Brassica rapa* L.) press cake and physicochemical properties of the protein fractions. *J. Agric. Food Chem.* **2015**, *63*, 2997–3003. [CrossRef] [PubMed]
29. Chirinos, R.; Aquino, M.; Pedreschi, R.; Campos, D. Optimized methodology for alkaline and enzyme assisted extraction of protein from Sacha Inchi (*Plukenetia volubilis*) kernel cake. *J. Food Proc. Eng.* **2016**, *40*, e12412. [CrossRef]
30. Houde, M.; Khodaei, N.; Benkerroum, N.; Karboune, S. Barley protein concentrates: Extraction, structural and functional properties. *Food Chem.* **2018**, *254*, 367–376. [CrossRef]
31. Perovic, M.N.; Zorica, D.K.J.; Mirjana, G.A. Improved recovery of protein from soy grit by enzyme-assisted alkaline extraction. *J. Food Eng.* **2020**, *276*, 109894. [CrossRef]
32. Naseri, A.; Marinho, G.S.; Holdt, S.L.; Bartela, J.M.; Jacobsen, C. Enzyme-assisted extraction and characterization of protein from red seaweed *Palmaria palmata*. *Algal Res.* **2020**, *47*, 101849. [CrossRef]
33. Akyüz, A.; Ersus, S. Optimization of enzyme assisted extraction of protein from the sugar beet (*Beta vulgaris* L.) leaves for alternative plant protein concentrate production. *Food Chem.* **2021**, *335*, 127673. [CrossRef]
34. De Souza, T.S.P.; Dias, F.F.G.; Oliveira, J.P.S.; de Moura Bell, J.M.N.L.; Koblitz, M.G.B. Biological properties of almond proteins produced by aqueous and enzyme-assisted aqueous extraction processes from almond cake. *Sci. Rep.* **2020**, *10*, 10873. [CrossRef]
35. Gil-Chávez, G.J.; Villa, J.A.; Ayala-Zavala, J.F.; Basilio Heredia, J.; Sepulveda, D.; Yahia, E.M.; González-Aguilar, G.A. Technologies for extraction and production of bioactive compounds to be used as nutraceuticals and food ingredients: An overview. *Compr. Rev. Food Sci. Food Saf.* **2013**, *12*, 5–23. [CrossRef]
36. Camel, V. Microwave-assisted solvent extraction of environmental samples. *Trends Anal. Chem.* **2000**, *19*, 229–248. [CrossRef]
37. Chan, C.H.; Yusoff, R.; Ngoh, G.C.; Kung, F.W.L. Microwave-assisted extractions of active ingredients from plants. *J. Chromatogr. A* **2011**, *1218*, 6213–6225. [CrossRef] [PubMed]
38. Chan, C.H.; Yusoff, R.; Ngoh, G.C. Assessment of scale-up parameters of microwave-assisted extraction via the extraction of flavonoids from cocoa leaves. *Chem. Eng. Technol.* **2015**, *38*, 489–496. [CrossRef]
39. Chemat, F.; Huma, Z.; Khan, M.K. Applications of ultrasound in food technology: Processing, preservation and extraction. *Ultrason. Sonochem.* **2011**, *18*, 813–835. [CrossRef]
40. Phongthai, S.; Lim, S.-T.; Rawdkuen, S. Optimization of microwave-assisted extraction of rice bran protein and its hydrolysates properties. *J. Cereal Sci.* **2016**, *70*, 154–156. [CrossRef]
41. Bedin, S.; Zanella, K.; Bragagnolo, N.; Taranto, O.P. Implications of microwaves on the extraction process of rice bran protein. *Braz. J. Chem. Eng.* **2019**, *36*, 1653–1665. [CrossRef]
42. Park, J.-H.; Choe, J.-H.; Kim, H.-W.; Hwang, K.-E.; Song, D.-H.; Yeo, E.-J.; Kim, H.-Y.; Choi, Y.-S.; Lee, S.-H.; Kim, C.-J. Effects of various extraction methods on quality characteristics of duck feet gelatin. *Korean J. Food Sci. An.* **2013**, *33*, 162–169. [CrossRef]
43. Zhu, K.-X.; Sun, X.-H.; Zhou, H.-M. Optimization of ultrasound assisted extraction of defatted wheat germ proteins by reverse micelles. *J. Cereal Sci.* **2009**, *50*, 266–271. [CrossRef]
44. Görgüç, A.; Bircan, C.; Yılmaz, F.M. Sesame bran as an unexploited by-product: Effect of enzyme and ultrasound assisted extraction on the recovery of protein and antioxidant compounds. *Food Chem.* **2019**, *283*, 637–645. [CrossRef]
45. Görgüç, A.; Özer, P.; Yılmaz, F.M. Microwave-assisted enzymatic extraction of plant protein with antioxidant compounds from the food waste sesame bran: Comparative optimization study and identification of metabolomics using LC/Q-OF/MS. *J. Food Process. Preserv.* **2019**, *44*, e14304. [CrossRef]
46. Leong, T.; Ashokkumar, M.; Kentish, S. The fundamentals of power ultrasound—A review. *Acoust. Aust.* **2011**, *39*, 54–63.
47. Rahman, M.M.; Lamsal, B.P. Ultrasound-assisted extraction and modification of plant-based proteins: Impact on physicochemical, functional, and nutritional properties. *Compr. Rev. Food Sci. Food Saf.* **2020**, *20*, 1457–1480. [CrossRef]
48. Bhargavaa, N.; Mor, R.S.; Kumar, K.; Sharanagat, V.J. Advances in application of ultrasound in food processing: A review. *Ultrason. Sonochem.* **2020**, *70*, 105293. [CrossRef]
49. Sun, X.; Zhang, W.; Zhang, L.; Tian, S.; Chen, F. Molecular and emulsifying properties of arachin and conarachin of peanut protein isolate from ultrasound-assisted extraction. *LWT-Food Sci. Technol.* **2020**, *132*, 109790. [CrossRef]
50. Wang, F.; Zhang, Y.; Xu, L.; Ma, H. An efficient ultrasound-assisted extraction method of pea protein and its effect on protein functional properties and biological activities. *LWT-Food Sci. Technol.* **2020**, *127*, 109348. [CrossRef]
51. Hadidi, M.; Baradaran Khaksar, F.; Pagan, J.; Ibarz, A. Application of ultrasound-ultrafiltration-assisted alkaline isoelectric precipitation (UUAAIP) technique for producing alfalfa protein isolate for human consumption: Optimization, comparison, physicochemical, and functional properties. *Food Res. Int.* **2020**, *130*, 108907. [CrossRef] [PubMed]
52. Zou, Y.; Li, P.P.; Zhang, K.; Wang, L.; Zhang, M.H.; Sun, Z.L.; Sun, C.; Geng, Z.M.; Xu, W.M.; Wang, D.Y. Effects of ultrasound-assisted alkaline extraction on the physiochemical and functional characteristics of chicken liver protein isolate. *Poult. Sci.* **2017**, *96*, 2975–2985. [CrossRef]

53. Mirzapour-Kouhdasht, A.; Sabzipour, F.; Taghizadeh, M.S.; Moosavi-Nasab, M. Physicochemical, rheological, and molecular characterization of colloidal gelatin produced from Common carp by-products using microwave and ultrasound-assisted extraction. *J. Texture Stud.* **2019**, *50*, 416–425. [CrossRef]
54. Pingret, D.; Fabiano-Tixier, A.S.; Chemat, F. Degradation during application of ultrasound in food processing: A review. *Food Cont.* **2013**, *31*, 593–606. [CrossRef]
55. Yang, X.; Li, Y.; Li, S.; Oladejo, A.O.; Wang, Y.; Huang, S.; Zhou, C.; Ye, X.; Ma, H.; Duan, Y. Effects of ultrasound-assisted α-amylase degradation treatment with multiple modes on the extraction of rice protein. *Ultrason. Sonochem.* **2018**, *40*, 890–899. [CrossRef] [PubMed]
56. Görgüç, A.; Özer, P.; Yılmaz, F.M. Simultaneous effect of vacuum and ultrasound assisted enzymatic extraction on the recovery of plant protein and bioactive compounds from sesame bran. *J. Food Comp. Anal.* **2020**, *87*, 103424. [CrossRef]
57. Li, W.; Yang, H.; Coldea, T.E.; Zhao, H. Modification of structural and functional characteristics of brewer's spent grain protein by ultrasound assisted extraction. *LWT-Food Sci. Technol.* **2021**, *139*, 110582. [CrossRef]
58. Varghese, T.; Pare, A. Effect of microwave assisted extraction on yield and protein characteristics of soymilk. *J. Food Eng.* **2019**, *262*, 92–99. [CrossRef]
59. Moreno-Nájera, L.C.; Ragazzo-Sánchez, J.A.; Gastón-Peña, C.R.; Calderón-Santoyo, M. Green technologies for the extraction of proteins from jackfruit leaves (*Artocarpus heterophyllus* Lam). *Food Sci. Biotechnol.* **2020**, *29*, 1675–1684. [CrossRef]
60. Elhag, H.E.E.A.; Naila, A.; Nour, A.H.; Ajit, A.; Sulaiman, A.Z.; Aziz, B.A. Optimization of protein yields by ultrasound assisted extraction from *Eurycoma longifolia* roots and effect of agitation speed. *J. King Saud Univ. Sci.* **2019**, *31*, 913–930. [CrossRef]
61. Wen, L.; Álvarez, C.; Zhang, Z.; Poojary, M.M.; Lund, M.N.; Sun, D.-W.; Tiwari, B.K. Optimisation and characterisation of protein extraction from coffee silverskin assisted by ultrasound or microwave techniques. *Biomass Convers. Biorefin.* **2020**. [CrossRef]
62. Zhao, Y.; Wen, C.; Feng, Y.; Zhang, J.; He, Y.; Duan, Y.; Zhang, H.; Ma, H. Effects of ultrasound-assisted extraction on the structural, functional and antioxidant properties of *Dolichos lablab* L. protein. *Proc. Biochem.* **2021**, *101*, 274–284. [CrossRef]
63. Tu, Z.-C.; Huang, T.; Wang, H.; Sha, X.-M.; Shi, Y.; Huang, X.-Q.; Man, Z.-Z.; Li, D.-J. Physico-chemical properties of gelatin from bighead carp (*Hypophthalmichthys nobilis*) scales by ultrasound-assisted extraction. *J. Food Sci. Technol.* **2015**, *52*, 2166–2174. [CrossRef]
64. Chemat, F.; Vian, M.A.; Fabiano-Tixier, A.S.; Nutrizio, M.; Jambrak, A.R.; Munekata, P.E.S.; Lorenzo, J.M.; Barba, F.J.; Binelloe, A.; Cravotto, G. A review of sustainable and intensified techniques for extraction of food and natural products. *Green Chem.* **2020**, *22*, 2325–2353. [CrossRef]
65. Ganeva, V.; Angelova, B.; Galutzov, B.; Goltsev, V.; Zhiponova, M. Extraction of proteins and other intracellular bioactive compounds from baker's yeasts by pulsed electric field treatment. *Front. Bioeng. Biotechnol.* **2020**, *8*, 552335. [CrossRef]
66. Arshad, R.N.; Abdul-Malek, Z.; Roobab, U.; Qureshi, M.I.; Khan, N.; Ahmad, M.H.; Liu, Z.-W.; Aadil, R.M. Effective valorization of food wastes and by-products through pulsed electric field: A systematic review. *J. Food Process. Eng.* **2021**, *44*, e13629. [CrossRef]
67. Martínez, J.M.; Delso, C.; Maza, M.A.; Raso, J. Utilising pulsed electric field processing to enhance extraction processes. In *Reference Module in Food Science*; Elsevier: New York, NY, USA, 2018. [CrossRef]
68. Roselló-Soto, E.; Barba, F.J.; Parniakov, O.; Galanakis, C.M.; Lebovka, N.; Grimi, N.; Vorobiev, E. High voltage electrical discharges, pulsed electric field, and ultrasound assisted extraction of protein and phenolic compounds from olive kernel. *Food Bioprocess. Technol.* **2014**. [CrossRef]
69. Sarkis, J.R.; Boussetta, N.; Blouet, C.; Tessaro, I.C.; Marczak, L.D.F.; Vorobiev, E. Effect of pulsed electric fields and high voltage electrical discharges on polyphenol and protein extraction from sesame cake. *Innov. Food Sci. Emerg. Technol.* **2015**, *29*, 170–177. [CrossRef]
70. 't Lama, G.P.; Postma, P.R.; Fernandes, D.A.; Timmermans, R.A.H.; Vermuë, M.H.; Barbosa, M.J.; Eppink, M.H.M.; Wijffels, R.H.; Olivieri, G. Pulsed electric field for protein release of the microalgae *Chlorella vulgaris* and *Neochloris oleoabundans*. *Algal Res.* **2017**, *24*, 181–187. [CrossRef]
71. Zhang, L.; Wang, L.-J.; Jiang, W.; Qian, J.-Y. Effect of pulsed electric field on functional and structural properties of canola protein by pretreating seeds to elevate oil yield. *Lebensm.-Wiss. Technol.* **2017**, *84*, 73–81. [CrossRef]
72. Liang, R.; Cheng, S.; Wang, X. Secondary structure changes induced by pulsed electric field affect antioxidant activity of pentapeptides from pine nut (*Pinus koraiensis*) protein. *Food Chem.* **2018**, *254*, 170–184. [CrossRef] [PubMed]
73. Shouqin, Z.; Jun, X.; Changzheng, W. High hydrostatic pressure extraction of flavonoids from propolis. *J. Chem. Technol. Biotechnol. Int. Res. Process Envir. Clean Technol.* **2005**, *80*, 50–54. [CrossRef]
74. Sezer, P.; Okur, I.; Oztop, M.H.; Alpas, H. Improving the physical properties of fish gelatin by high hydrostatic pressure (HHP) and ultrasonication (US). *Int. J. Food Sci. Technol.* **2019**, *55*, 1468–1476. [CrossRef]
75. Suwal, S.; Perreault, V.; Marciniak, A.; Tamigneaux, E.; Deslandes, E.; Bazinet, L.; Jacques, H.; Beaulieu, L.; Doyen, A. Effects of high hydrostatic pressure and polysaccharidases on the extraction of antioxidant compounds from red macroalgae, *Palmaria palmata Solieria chordalis*. *J. Food Eng.* **2019**, *52*, 53–59. [CrossRef]
76. Cascaes Teles, A.S.; Hidalgo Chávez, D.W.; Zarur Coelho, M.A.; Rosenthal, A.; Fortes Gottschalk, L.M.; Tonon, R.V. Combination of enzyme-assisted extraction and high hydrostatic pressure for phenolic compounds recovery from grape pomace. *J. Food Eng.* **2020**, *288*, 110128. [CrossRef]

77. Bolat, B.; Ugur, A.E.; Oztop, M.H.; Alpas, H. Effects of high hydrostatic pressure assisted degreasing on the technological properties of insect powders obtained from *Acheta domesticus* & *Tenebrio molitor*. *J. Food Eng.* **2021**, *292*, 110359. [CrossRef]

78. Gharibzahedi, S.M.T.; Smith, B. Effects of high hydrostatic pressure on the quality and functionality of protein isolates, concentrates, and hydrolysates derived from pulse legumes: A review. *Trends Food Sci. Technol.* **2021**, *107*, 466–479. [CrossRef]

79. Marciniak, A.; Suwal, S.; Naderi, N.; Pouliot, Y.; Doyen, A. Enhancing enzymatic hydrolysis of food proteins and production of bioactive peptides using high hydrostatic pressure technology. *Trends Food Sci. Technol.* **2018**, *80*, 187–198. [CrossRef]

80. Rivalain, N.; Roquain, J.; Demazeau, G. Development of high hydrostatic pressure in biosciences: Pressure effect on biological structures and potential applications in Biotechnologies. *Biotechnol. Adv.* **2010**, *28*, 659–672. [CrossRef]

81. Ahmed, J.; Mulla, M.; Al-Ruwaih, N.; Arfat, Y.A. Effect of high-pressure treatment prior to enzymatic hydrolysis on rheological, thermal, and antioxidant properties of lentil protein isolate. *Legume Sci.* **2019**, *1*, e10. [CrossRef]

82. Tribst, A.A.L.; Ribeiro, L.R.; Cristianini, M. Comparison of the effects of high pressure homogenization and high pressure processing on the enzyme activity and antimicrobial profile of lysozyme. *Innov. Food Sci. Emerg. Technol.* **2017**, *43*, 60–67. [CrossRef]

83. Ulug, S.K.; Jahandideh, F.; Wu, J. Novel technologies for the production of bioactive peptides. *Trends Food Sci. Technol.* **2021**, *108*, 27–39. [CrossRef]

84. Espinoza, A.D.; Morawicki, R.O.; Hager, T. Hydrolysis of whey protein isolate using subcritical water. *J. Food Sci.* **2012**, *71*, C20–C26. [CrossRef] [PubMed]

85. Hall, F.; Liceaga, A. Effect of microwave-assisted enzymatic hydrolysis of cricket (*Gryllodes sigillatus*) protein on ACE and DPP-IV inhibition and tropomyosin-IgG binding. *J. Funct. Foods* **2020**, *64*, 103634. [CrossRef]

86. Urbizo-Reyes, U.; San Martin-González, M.F.; Garcia-Bravo, J.; López Malo Vigil, A.; Liceaga, A.M. Physicochemical characteristics of chia seed (*Salvia hispanica*) protein hydrolysates produced using ultrasonication followed by microwave-assisted hydrolysis. *Food Hydrocoll.* **2019**, *97*, 105187. [CrossRef]

87. Chen, Z.; Li, Y.; Lin, S.; Wei, M.; Du, F.; Ruan, G. Development of continuous microwave-assisted protein digestion with immobilized enzyme. *Biochem. Biophys. Res. Commun.* **2014**, *445*, 491–496. [CrossRef] [PubMed]

88. Uluko, H.; Zhang, S.; Liu, L.; Tsakama, M.; Lu, J.; Lv, J. Effects of thermal, microwave, and ultrasound pretreatments on antioxidative capacity of enzymatic milk protein concentrate hydrolysates. *J. Funct. Foods* **2015**, *18*, 1138–1146. [CrossRef]

89. Yang, F.; Hu, F.; Jiang, Q.; Xu, Y.; Xia, W. Effect of pretreatments on hydrolysis efficiency and antioxidative activity of hydrolysates produced from bighead carp (*Aristichthys nobilis*). *J. Aquat. Food Prod. Technol.* **2016**, *25*, 916–927. [CrossRef]

90. Nguyen, E.; Jones, O.; Kim, Y.H.B.; San Martin-Gonzalez, F.; Liceaga, A.M. Impact of microwave-assisted enzymatic hydrolysis on functional and antioxidant properties of rainbow trout *Oncorhynchus mykiss* by-products. *Fish. Sci.* **2017**, *83*, 317–331. [CrossRef]

91. Ketnawa, S.; Wickramathilaka, M.; Liceaga, A.M. Changes on antioxidant activity of microwave-treated protein hydrolysates after simulated gastrointestinal digestion: Purification and identification. *Food Chem.* **2018**, *254*, 36–46. [CrossRef]

92. Li, Y.; Li, J.; Lin, S.J.; Yang, Z.S.; Jin, H.X. Preparation of antioxidant peptide by microwave-assisted hydrolysis of collagen and its protective effect against H_2O_2-induced damage of RAW264.7 cells. *Mar. Drugs* **2019**, *17*, 642. [CrossRef]

93. Sparr Eskilsson, C.; Björklund, E. Analytical-scale microwave-assisted extraction. *J. Chromatogr. A* **2000**, *902*, 227–250. [CrossRef]

94. Jia, J.; Ma, H.; Zhao, W.; Wang, Z.; Tian, W.; Luo, L.; He, R. The use of ultrasound for enzymatic preparation of ACE-inhibitory peptides from wheat germ protein. *Food Chem.* **2010**, *119*, 336–342. [CrossRef]

95. Zhu, K.-X.; Su, C.-Y.; Guo, X.-N.; Peng, W.; Zhou, H.-M. Influence of ultrasound during wheat gluten hydrolysis on the antioxidant activities of the resulting hydrolysate. *Int. J. Food Sci. Technol.* **2011**, *46*, 1053–1059. [CrossRef]

96. Wali, A.; Ma, H.; Shahnawaz, M.; Hayat, K.; Xiaong, J.; Jing, L. Impact of power ultrasound on antihypertensive activity, functional properties, and thermal stability of rapeseed protein hydrolysates. *J. Chem.* **2017**, *2017*, 4373859. [CrossRef]

97. Wu, Q.; Zhang, X.; Jia, J.; Kuang, C.; Yang, H. Effect of ultrasonic pretreatment on whey protein hydrolysis by alcalase: Thermodynamic parameters, physicochemical properties and bioactivities. *Proc. Biochem.* **2018**, *67*, 46–54. [CrossRef]

98. Lei, B.; Majumder, K.; Shen, S.; Wu, J. Effect of sonication on thermolysin hydrolysis of ovotransferrin. *Food Chem.* **2011**, *124*, 808–815. [CrossRef]

99. Piccolomini, A.; Iskandar, M.; Lands, L.; Kubow, S. High hydrostatic pressure pre-treatment of whey proteins enhances whey protein hydrolysate inhibition of oxidative stress and IL-8 secretion in intestinal epithelial cells. *Food Nutr. Res.* **2012**, *56*, 17549. [CrossRef]

100. Boukil, A.; Suwal, S.; Chamberland, J.; Pouliot, Y.; Doyen, A. Ultrafiltration performance and recovery of bioactive peptides after fractionation of tryptic hydrolysate generated from pressure-treated B-lactoglobulin. *J. Membr. Sci.* **2018**, *556*, 42–53. [CrossRef]

101. Garcia-Mora, P.; Peñas, E.; Frias, J.; Zielinski, H.; Wiczkowski, W.; Zielinska, D.; Martínez-Villaluenga, C. High-pressure-assisted enzymatic release of peptides and phenolics increases angiotensin converting enzyme I inhibitory and antioxidant activities of pinto bean hydrolysates. *J. Agric. Food Chem.* **2016**, *64*, 1730–1740. [CrossRef]

102. Al-Ruwaih, N.; Ahmed, J.; Mulla, M.F.; Arfat, Y.A. High-pressure assisted enzymatic proteolysis of kidney beans protein isolates and characterization of hydrolysates by functional, structural, rheological and antioxidant properties. *LWT- Food Sci. Technol.* **2019**, *100*, 231–236. [CrossRef]

103. Guan, H.; Diao, X.; Jiang, F.; Han, J.; Kong, B. The enzymatic hydrolysis of soy protein isolate by corolase PP under high hydrostatic pressure and its effect on bioactivity and characteristics of hydrolysates. *Food Chem.* **2018**, *245*, 89–96. [CrossRef]

104. Perreault, V.; Hénaux, L.; Bazinet, L.; Doyen, A. Pretreatment of flaxseed protein isolate by high hydrostatic pressure: Impacts on protein structure, enzymatic hydrolysis and final hydrolysate antioxidant capacities. *Food Chem.* **2017**, *221*, 1805–1812. [CrossRef]

105. Franco, D.; Munekata, P.E.S.; Agregán, R.; Bermúdez, R.; López-Pedrouso, M.; Pateiro, M.; Lorenzo, J.M. Application of pulsed electric fields for obtaining antioxidant extracts from fish residues. *Antioxidants* **2020**, *9*, 90. [CrossRef] [PubMed]

106. Lin, S.; Guo, Y.; Liu, J.; You, Q.; Yin, Y.; Cheng, S. Optimized enzymatic hydrolysis and pulsed electric field treatment for production of antioxidant peptides from egg white protein. *Afr. J. Biotechnol.* **2011**, *10*, 11648–11657.

107. Lin, S.; Guo, Y.; You, Q.; Yin, Y.; Liu, J. Preparation of antioxidant peptide from egg white protein and improvement of its activities assisted by high-intensity pulsed electric field. *J. Sci. Food Agric.* **2012**, *92*, 1554–1561. [CrossRef] [PubMed]

108. Lin, S.; Jin, Y.; Liu, M.; Yang, Y.; Zhang, M.; Guo, Y.; Jones, G.; Liu, J.; Yin, Y. Research on the preparation of antioxidant peptides derived from egg white with assisting of high intensity pulsed electric field. *Food Chem.* **2013**, *139*, 300–306. [CrossRef]

109. Lin, S.; Liang, R.; Li, X.; Xing, J.; Yuan, Y. Effect of pulsed electric field (PEF) on structures and antioxidant activity of soybean source peptides-SHCMN. *Food Chem.* **2016**, *213*, 588–594. [CrossRef] [PubMed]

110. Kumar, Y.; Patel, K.K.; Kumar, V. Pulsed electric field processing in food technology. *Int. J. Eng. Stud. Techn. Approach* **2015**, *1*, 6–16. [CrossRef]

111. Asaduzzaman, A.K.M.; Getachewa, A.T.; Cho, Y.-J.; Park, J.S.; Haq, M.; Chun, B.-S. Characterization of pepsin-solubilised collagen recovered from mackerel (*Scomber japonicus*) bone and skin using subcritical water hydrolysis. *Int. J. Biol. Macromol.* **2020**, *148*, 1290–1297. [CrossRef]

112. Wang, M.P.; Lu, W.; Yang, J.; Wang, J.M.; Yang, X.Q. Preparation and characterisation of isoflavone aglycone-rich calcium-binding soy protein hydrolysates. *Int. J. Food Sci. Technol.* **2017**, *52*, 2230–2237. [CrossRef]

113. Han, J.; Kim, M.R.; Park, Y.; Hong, Y.H.; Suh, H.J. Skin permeability of porcine placenta extracts and its physiological activities. *Korean J. Food Sci. Anim. Res.* **2013**, *33*, 356–362. [CrossRef]

114. Espinosa-Pardo, F.A.; Savoire, R.; Subra-Paternault, P.; Harscoat-Schiavo, C. Oil and protein recovery from corn germ: Extraction yield, composition and protein functionality. *Food Bioprod. Process.* **2020**, *120*, 131–142. [CrossRef]

115. Li, Y.; Shi, J.; Scanlon, M.; Xue, S.J.; Lu, J. Effects of pretreatments on physicochemical and structural properties of proteins isolated from canola seeds after oil extraction by supercritical-CO_2 process. *LWT Food Sci. Technol.* **2020**, *137*, 110415. [CrossRef]

116. Olivera-Montenegro, L.; Best, I.; Gil-Saldarriaga, A. Effect of pretreatment by supercritical fluids on antioxidant activity of protein hydrolyzate from quinoa (*Chenopodium quinoa* Willd.). *Food Sci. Nutr.* **2021**, *9*, 574–582. [CrossRef]

117. Koh, B.-B.; Lee, E.-J.; Ramachandraiah, K.; Hong, G.-P. Characterization of bovine serum albumin hydrolysates prepared by subcritical water processing. *Food Chem.* **2019**, *278*, 203–207. [CrossRef]

118. Powell, T.; Bowra, S.; Cooper, H.J. Subcritical water processing of proteins: An alternative to enzymatic digestion. *Anal. Chem.* **2016**, *88*, 6425–6432. [CrossRef]

119. Choi, J.-S.; Moon, H.E.; Roh, M.K.; Ha, V.M.; Lee, B.B.; Cho, K.K.; Choi, I.S. Physiological properties of Scomber japonicus meat hydrolysate prepared by subcritical water hydrolysis. *J. Env. Biol.* **2016**, *37*, 57–63.

120. Haq, M.; Ho, T.C.; Ahmed, R.; Getachew, A.T.; Cho, Y.-J.; Park, J.-S.; Chun, B.-S. Biofunctional properties of bacterial collagenolytic protease-extracted collagen hydrolysates obtained using catalysts-assisted subcritical water hydrolysis. *J. Ind. Eng. Chem.* **2020**, *81*, 332–339. [CrossRef]

121. Ahmed, R.; Chun, B.S. Subcritical water hydrolysis for the production of bioactive peptides from tuna skin collagen. *J. Supercrit. Fluids* **2018**, *141*, 88–96. [CrossRef]

122. Melgosa, R.; Marques, M.; Paiva, A.; Bernardo, A.; Fernández, N.; Sá-Nogueira, I.; Simões, P. Subcritical water extraction and hydrolysis of cod (*Gadus morhua*) frames to produce bioactive protein extracts. *Foods* **2021**, *10*, 1222. [CrossRef]

123. Lee, H.-J.; Roy, V.C.; Ho, T.C.; Park, J.-S.; Jeong, Y.-R.; Lee, S.C.; Kim, S.-Y.; Chun, B.-S. Amino acid profiles and biopotentiality of hydrolysates obtained from comb penshell (*Atrina pectinata*) viscera using subcritical water hydrolysis. *Mar. Drugs* **2021**, *19*, 137. [CrossRef] [PubMed]

124. Polikovsky, M.; Gillis, A.; Steinbruch, E.; Robin, A.; Epstein, M.; Kribus, A.; Golberg, A. Biorefinery for the co-production of protein, hydrochar and additional co-products from a green seaweed *Ulva* sp. with subcritical water hydrolysis. *Energy Convers. Manag.* **2020**, *225*, 113380. [CrossRef]

125. Park, J.-S.; Jeong, Y.-R.; Chun, B.S. Physiological activities and bioactive compound from laver (*Pyropia yezoensis*) hydrolysates by using subcritical water hydrolysis. *J. Supercrit. Fluids* **2019**, *148*, 130–136. [CrossRef]

126. Ramachandraiah, K.; Koh, B.-B.; Davaatseren, M.; Hong, G.-P. Characterization of soy protein hydrolysates produced by varying subcritical water processing temperature. *Innov. Food Sci. Emerg. Technol.* **2017**, *43*, 201–206. [CrossRef]

127. Fusaro, F.; Kluge, J.; Mazzotti, M.; Muhrer, G. Compressed CO_2 antisolvent precipitation of lysozyme. *J. Supercrit. Fluids* **2009**, *49*, 79–92. [CrossRef]

128. Moshashaée, S.; Bisrat, M.; Forbes, R.T.; Nyqvist, H.; York, P. Supercritical fluid processing of proteins I: Lysozyme precipitation from organic solution. *Eur. J. Pharm. Sci.* **2000**, *11*, 239–245. [CrossRef]

129. Yver, A.L.; Bonnaillie, L.M.; Yee, W.; McAloon, A.; Tomasula, P.M. Fractionation of whey protein isolate with supercritical carbon dioxide — Process modeling and cost estimation. *Int. J. Mol. Sci.* **2012**, *13*, 240–259. [CrossRef]

130. Perinelli, D.R.; Bonacucina, G.; Cespi, M.; Naylor, A.; Whitaker, M.; Palmieri, G.F.; Giorgioni, G.; Casettari, L. Evaluation of P(L)LA-PEG-P(L)LA as processing aid for biodegradable particles from gas saturated solutions (PGSS) process. *Int. J. Pharm.* **2014**, *468*, 250–257. [CrossRef]

131. Zhong, Q.; Jin, M.; Xiao, D.; Tian, H.; Zhang, W. Application of supercritical antisolvent technologies for the synthesis of delivery systems of bioactive food components. *Food Biophys.* **2008**, *3*, 186–190. [CrossRef]
132. Kluge, J.; Fusaro, F.; Casas, N.; Mazzotti, M.; Muhrer, G. Production of PLGA micro and nanocomposites by supercritical fluid extraction of emulsions: I. encapsulation of lysozyme. *J. Supercrit. Fluids* **2009**, *50*, 327–335. [CrossRef]
133. Reibe, C.; Knez, Z.; Weidner, E. A new high-pressure micronisation process for the gentle processing of high molecular mass gelatine. *Food Bioprod. Proc.* **2012**, *90*, 79–86. [CrossRef]
134. Nuchuchua, O.; Every, H.A.; Hofland, G.W.; Jiskoot, W. Scalable organic solvent free supercritical fluid spray drying process for producing dry protein formulations. *Eur. J. Pharm. Biopharm.* **2014**, *88*, 919–930. [CrossRef] [PubMed]
135. Bouchard, A.; Jovanovic, N.; Jiskoot, W.; Mendes, E.; Witkamp, G.-J.; Crommelin, D.J.A.; Hofland, G.W. Lysozyme particle formation during supercritical fluid drying: Particle morphology and molecular integrity. *J. Supercrit. Fluids* **2007**, *40*, 293–307. [CrossRef]
136. De Marco, I.; Reverchon, E. Supercritical carbon dioxide plus ethanol mixtures for the antisolvent micronization of hydrosoluble materials. *Chem. Eng. J.* **2012**, *187*, 401–409. [CrossRef]
137. Adami, R.; Osseo, L.S.; Reverchon, E. Micronization of lysozyme by supercritical assisted atomization. *Biotechnol. Bioeng.* **2009**, *104*, 1162–1170. [CrossRef]
138. Wang, Q.; Guan, Y.-X.; Yao, S.-J.; Zhu, Z.-Q. Controllable preparation and formation mechanism of BSA microparticles using supercritical assisted atomization with an enhanced mixer. *J. Supercrit. Fluids* **2011**, *56*, 97–104. [CrossRef]
139. Du, Z.; Guan, Y.-X.; Yao, S.-J.; Zhu, Z.-Q. Supercritical fluid assisted atomization introduced by an enhanced mixer for micronization of lysozyme: Particle morphology, size and protein stability. *Int. J. Pharm.* **2011**, *421*, 258–268. [CrossRef] [PubMed]
140. Shen, Y.-B.; Cuan, Y.-X.; Yao, S.-J. Supercritical fluid assisted production of micrometric powders of the labile trypsin and chitosan/trypsin composite microparticles. *Int. J. Pharm.* **2015**, *489*, 226–236. [CrossRef]
141. Trivedi, V.; Bhomia, R.; Mitchell, J.C. Myristic acid coated protein immobilised mesoporous silica particles as pH induced oral delivery system for the delivery of biomolecules. *Pharm.* **2019**, *12*, 153. [CrossRef]
142. Gañan, N.; Bordón, M.G.; Ribotta, P.D.; González, A. Study of chia oil microencapsulation in soy protein microparticles using supercritical CO_2-assisted impregnation. *J. CO_2 Util.* **2020**, *40*, 101221. [CrossRef]
143. Lima, J.C.; Seixas, F.A.V.; Coimbra, J.S.R.; Pimentel, T.C.; Barão, C.E.; Cardozo-Filho, L. Continuous fractionation of whey protein isolates by using supercritical carbon dioxide. *J. CO_2 Util.* **2019**, *30*, 112–122. [CrossRef]
144. Lima, J.C.; Bonfim-Rocha, L.; Barao, C.E.; Coimbra, J.S.R.; Cardozo-Filho, L. Techno-Economic assessment of α-Lactalbumin and β-Lactoglobulin fractionation from whey protein isolated solution using supercritical carbon dioxide in a continuous reactor. *J. Taiwan Inst. Chem. Eng.* **2021**, *118*, 87–96. [CrossRef]
145. Martin, A.; Weidner, E. PGSS-drying: Mechanisms and modeling. *J. Supercrit. Fluids* **2010**, *55*, 271–281. [CrossRef]
146. Seifried, B.; Temelli, F. Supercritical fluid drying of high molecular weight biopolymers for particle formation and delivery of bioactives. In Proceedings of the 10th International Symposium Supercritical Fluids (ISSF2012), San Francisco, CA, USA, 13–16 May 2012.
147. Rodrigues, M.A.; Li, J.; Padrela, L.; Almeida, A.; Matos, H.A.; de Azevedo, E.G. Anti-solvent effect in the production of lysozyme nanoparticles by supercritical fluid-assisted atomization processes. *J. Supercrit. Fluids.* **2009**, *48*, 253–260. [CrossRef]
148. Dickinson, E. Milk protein interfacial layers and the relationship to emulsion stability and rheology. *Colloids Surf. B Biointerfaces* **2001**, *20*, 197–210. [CrossRef]
149. Liu, X.; Powers, J.R.; Swanson, B.G.; Hill, H.H.; Clark, S. Modification of whey protein concentrate hydrophobicity by high hydrostatic pressure. *Innov. Food Sci. Emerg. Technol.* **2005**, *6*, 310–317. [CrossRef]
150. Xu, D.; Yuan, F.; Jiang, J.; Wang, X.; Hou, Z.; Gao, Y. Structural and conformational modification of whey proteins induced by supercritical carbon dioxide. *Innov. Food Sci. Emerg. Technol.* **2011**, *12*, 32–37. [CrossRef]
151. Manoi, K.; Rizvi, S.S.H. Rheological characterization of texturized whey protein concentrate-based powders produced by reactive supercritical fluid extrusion. *Food Res. Int.* **2008**, *41*, 786–796. [CrossRef]
152. Liu, X.; Gao, Y.; Xu, H.; Wang, Q.; Yang, B. Impact of high-pressure carbon dioxide combined with thermal treatment on degradation of red beet (*Beta vulgaris* L.) pigments. *J. Agric. Food Chem.* **2008**, *56*, 6480–6487. [CrossRef]
153. Yin, C.Y.; Li, J.B.; Xu, Q.; Peng, Q.; Liu, Y.B.; Shen, X.Y. Chemical modification of cotton cellulose in supercritical carbon dioxide: Synthesis and characterization of cellulose carbamate. *Carbohydr. Polym.* **2007**, *67*, 147–154. [CrossRef]
154. Zhong, Q.; Jin, M. Enhanced functionalities of whey proteins treated with supercritical carbon dioxide. *J. Dairy Sci.* **2008**, *91*, 490–499. [CrossRef] [PubMed]
155. Monhemi, H.; Dolatabadi, S. Molecular dynamics simulation of high-pressure CO_2 pasteurization reveals the interfacial denaturation of proteins at CO_2/water interface. *J. CO_2 Util.* **2020**, *35*, 256–264. [CrossRef]
156. Monhemi, H.; Housaindokht, M.R. The molecular mechanism of protein denaturation in supercritical CO_2: The role of exposed lysine residues is explored. *J. Supercrit. Fluids* **2019**, *147*, 222–230. [CrossRef]
157. Vatansever, S.; Rao, J.; Hall, C. Effects of ethanol modified supercritical carbon dioxide extraction and particle size on the physical, chemical, and functional properties of yellow pea flour. *Cereal Chem.* **2020**, *97*, 1133–1147. [CrossRef]
158. Podrepsek Yoon, A.K.; Singha, P.; Rizvi, S.S.H. Steam vs. SC–CO_2–based extrusion: Comparison of physical properties of milk protein concentrate extrudates. *J. Food Eng.* **2021**, *292*, 110244. [CrossRef]

159. Gopirajah, R.; Singha, P.; Javad, S. Emulsifying properties of milk protein concentrate functionalized by supercritical fluid extrusion. *J. Food Proc. Preserv.* **2020**, *44*, e14754. [CrossRef]
160. Ding, L.; Zhao, Q.; Zhou, X.; Tang, C.; Chen, Y.; Cai, Z. Changes in protein structure and physicochemical properties of egg white by super critical carbon dioxide treatment. *J. Food Eng.* **2020**, *284*, 110076. [CrossRef]

Review

Extraction Methods of Oils and Phytochemicals from Seeds and Their Environmental and Economic Impacts

Valerie M. Lavenburg [1], Kurt A. Rosentrater [2] and Stephanie Jung [1,*]

[1] Food Science and Nutrition Department, California Polytechnic State University, San Luis Obispo, CA 93407, USA; vlavenbu@calpoly.edu
[2] Department of Agricultural and Biosystems Engineering, Iowa State University, Ames, IA 50011, USA; karosent@iastate.edu
* Correspondence: stjung@calpoly.edu

Abstract: Over recent years, the food industry has striven to reduce waste, mostly because of rising awareness of the detrimental environmental impacts of food waste. While the edible oils market (mostly represented by soybean oil) is forecasted to reach 632 million tons by 2022, there is increasing interest to produce non-soybean, plant-based oils including, but not limited to, coconut, flaxseed and hemp seed. Expeller pressing and organic solvent extractions are common methods for oil extraction in the food industry. However, these two methods come with some concerns, such as lower yields for expeller pressing and environmental concerns for organic solvents. Meanwhile, supercritical CO_2 and enzyme-assisted extractions are recognized as green alternatives, but their practicality and economic feasibility are questioned. Finding the right balance between oil extraction and phytochemical yields and environmental and economic impacts is challenging. This review explores the advantages and disadvantages of various extraction methods from an economic, environmental and practical standpoint. The novelty of this work is how it emphasizes the valorization of seed by-products, as well as the discussion on life cycle, environmental and techno-economic analyses of oil extraction methods.

Keywords: seeds; supercritical CO_2 extraction; solvent extraction; expeller pressing; enzyme-assisted aqueous extraction; techno-economic analysis; life cycle assessment

Citation: Lavenburg, V.M.; Rosentrater, K.A.; Jung, S. Extraction Methods of Oils and Phytochemicals from Seeds and Their Environmental and Economic Impacts. *Processes* **2021**, *9*, 1839. https://doi.org/10.3390/pr9101839

Academic Editors: Juliana Maria Leite Nobrega De Moura Bell, Blanca Hernández-Ledesma and Roberta Claro da Silva

Received: 28 September 2021
Accepted: 12 October 2021
Published: 16 October 2021

Publisher's Note: MDPI stays neutral with regard to jurisdictional claims in published maps and institutional affiliations.

1. Introduction

The sustainability and valorization of by-products have become an important focus of the food industry over the past few years. According to the Food and Agriculture Organization (FAO) of the United Nations, every year, approximately 1.3 billion tons, equivalent to 30% of total food production, is wasted globally. This volume of food waste is worth USD 750 million. Several initiatives have been implemented to combat food waste. In 2013, the United States Department of Agriculture (USDA) partnered with the United States Environmental Protection Agency (US EPA) to formally set a goal to reduce the country's food waste by 50% by 2030 [1,2]. Additionally, the EPA identified a food waste hierarchy that prioritizes feeding hungry people, feeding animals, industrial use, composting, then incineration or landfilling (in order of decreasing preference) (Figure 1) [3]. The Food Recovery Act of 2017 instituted various guidelines encouraging farms, grocery stores, restaurants and institutions to donate excess food, set up composting and anaerobic digestion programs and reduce overall food waste [4]. In California, the legislation "Senate Bill 1383" requires businesses to recover at least 20% of disposed edible food and divert it for human consumption by 2025 [5].

Among the various types of food waste generated, roots, tubers and fruits and vegetables are the most notable ones, representing 45% of the total waste. In addition, a whopping 20% of oilseeds, which come from crops such as sunflowers grown specifically to produce edible oil, are lost during agricultural production and postharvest handling and storage [6,7]. This is a tremendous amount, considering that global oilseed production was forecasted to reach 632 million tons during 2021–2022, and is expected to be worth USD 162.5 billion by 2025 [8,9]. Furthermore, processors are aiming to reduce all forms of seed-related waste by applying various strategies including but not limited to valorizing by-products by extracting residual phytochemicals or oil for the food, cosmetic or pharmaceutical industries. For example, tomato seeds, recovered during the processing of tomato-based products such as paste and ketchup, could be a source of edible oil. Meanwhile, in recent years consumer preference has grown for foods that promote health benefits, are environmentally friendly and offer a pleasant taste and aroma. As a result, the market for specialty oils (which refer to non-commodity oils with functional properties that are not further refined, bleached, or deodorized) has considerably increased [10]. Seeds indeed often contain desired unsaturated fatty acids and phytochemical components, which exhibit antioxidant and anti-cancer effects [11]. Among the specialty oils, coconut and olive oils (the latter of which is not from an oilseed) have become popular, with production reaching 3.67 million tons and 3.1 million tons, respectively, in 2020 [12]. Other less popular specialty oils include sesame, flaxseed and hempseed oils [13]. It should be noted that identifying what crops are grown specifically for their oilseeds (as opposed to crops that have seeds but are also utilized for other purposes) can sometimes be confusing. Thus, a summary of the categories of all the matrices discussed in this paper is included for the purpose of clarity for the readers (Table 1).

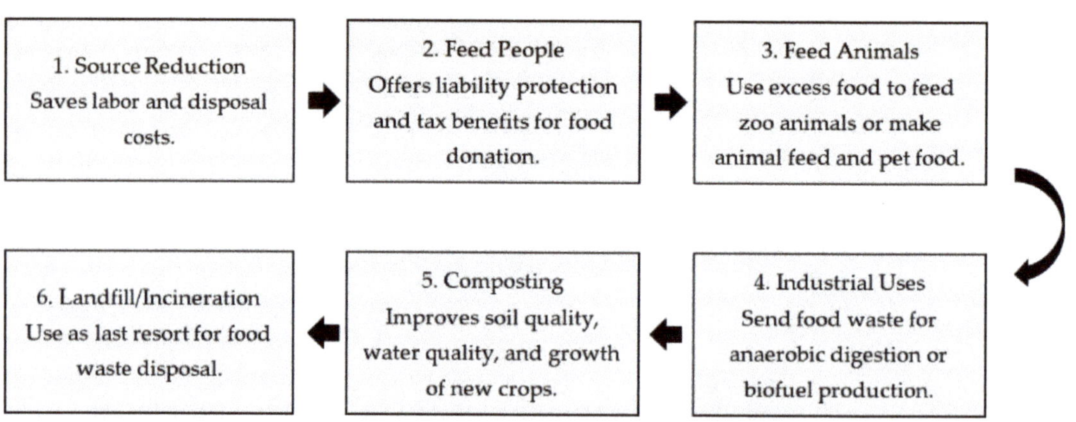

Figure 1. Priorities of the food recovery hierarchy, from most desirable (1) to least desirable (6). Adapted from [3].

Table 1. Types of extracted seed matrices in research studies covered in this literature review.

Oilseed	Other Seed Type
Canola/Rapeseed	Almond
Linseed (also known as flaxseed)	Coriander
Jatropha curcas	Hemp
Soybean	Favela
Sunflower	Passionfruit
Camelina	Peach
Castor	*Elaeagnus mollis*
Mustard	Moringa
Peanut	Grape
	Forsythia suspense

Recently, the food industry has prioritized balancing the economic and environmental aspects of edible oil production. This shift has been driven mainly by the consumers who have been more conscientious about sustainable food production and its three pillars:

people, planet and profit. Among the oil extraction methods being considered, expeller pressing and solvent extractions are most commonly used at an industrial scale [14]. However, both methods have some major pitfalls to overcome: a lower oil yield with expeller pressing compared to solvent extraction can make the process economically disadvantageous, and the use of organic solvents brings environmental concerns. With the increasing focus on the environmental impact of unit operations used during processing and the development of green chemistry, more studies have focused on improving the extraction methods so that less energy is required and less chemical pollutants are released by these processes [15]. Supercritical CO_2 (SCO_2) and enzyme-assisted extractions are alternatives to solvent extraction and expeller pressing, which are considered traditional oil extraction methods.

This review includes studies on oil extractions of seeds that were published between 2010 until the present day, with the exception of a few studies that are older. The purpose of this review is to explore the advantages and disadvantages of oil extraction methods of seeds from the lens of sustainability and food waste reduction, as well as life cycle, environmental impact and techno-economic analyses.

2. Mechanical Pressing

Historically, oil has been pressed out of seeds by indigenous communities for centuries, and the mechanical pressing of soybeans dates back to the 1940s [14]. There are two broad categories of equipment for oil extraction: expeller press and extruder (Figure 2). Expeller pressing is often limited to small scale, on-farm seed grinding operations. For example, canola, sunflower, flax and safflower oils are extracted via expeller pressing in the mid-west and northeastern United States. Due to its low cost, expeller pressing is also often used in developing countries, such as rural India, for linseed oil extraction [14,16].

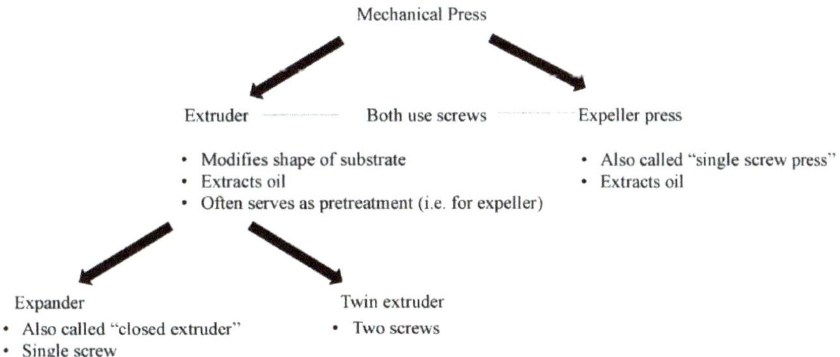

Figure 2. Major types of mechanical pressing.

An expeller press has a screw that rotates in a perforated barrel. The discharge area is partially obstructed, exerting pressure onto seeds to extract oil. Expeller pressing is considered an easy method for oil extraction because it only requires mechanical power and does not need organic solvents [17]. The extraction temperature can be kept under 50 °C to perform cold pressing, which can help preserve nutritional compounds of the oil [18]. However, one disadvantage of this method can be its lower oil recovery. If spacing is too small within the perforated barrel, or if high compaction of seeds results from pressing, it can jam the operating screw and leave 5 to 20% of the total oil in the residual cake [17,19]. Thus, there has been interest in making expeller pressing more efficient and, therefore, more economically viable.

Several parameters need to be considered to improve oil extraction yield and quality, and one such example is screw rotation. When using a pilot expeller press designed for cold pressing, increasing rotation from 1.2 to 18 rpm increased press capacity from 2.2 kg seed/h to 29.4 seed/h, while decreasing canola oil yields from 91 to 84% [20]. Additionally, the number of presses was shown to affect the oil yield and the quality of linseed oil. Increasing the number of presses of an expeller press from one to two increased linseed oil yields from 19 to 32%, while adding a third press did not significantly affect the oil yield. Implementing a double press also led to the highest total phenol (27 mg GAE/100 g, which was a 170% increase from a single press) and total flavonoid content (7 mg rutin eq/100 eq, which was a 40% increase from a single press). Applying more than two presses started to degrade these compounds due to the high pressure and temperature [16]. Another approach to increase the oil yield was to blend oilseeds to improve the consistency of the matrix, which enhanced the permeability and oil recovery. For example, an oil recovery of 94.7% was obtained with *Jatropha* seeds blended with soybean, and decreased to 88.4 and 75.4% when blended with maize and rapeseed, respectively [21].

Another way to improve oil extraction is to perform extrusion of the seeds (Figure 2). This process, which also relies on screw configuration, is used to modify the shape and properties in applications such as expanded snacks (such as cheerios) or obtain liquid extracts from plant material. The end of the screw allows for seeds to be extruded through a perforated plate and discharges oil [17,22,23]. Extrusion has been used as pretreatment prior to expeller pressing of soybean oil, extracting over 70% of oil compared to single-step expelling, which yielded 60% [24]. This process has also been used for simultaneous treatment with fatty acid methyl ester as a solvent, extracting 98% of oil from sunflower seeds [25].

Single-screw extrusion (expander) is mostly employed at a large-scale for the pressing of oil from seeds, but twin screw systems are used in laboratory and pilot studies [18,23]. The advantage of a twin extruder is that it allows for a thermomechanical treatment of seeds and avoids further pre-treatment steps (such as dehulling, flaking, cooking) often necessary to obtain high oil yields from single-screw operations [17,22,23]. A twin extruder set to 50 rpm and a flow rate of 2.27 kg/h can extract up to 50% of coriander oil without any pretreatment [26]. Fifty percent of oil was also obtained from sunflower seeds with parameters set to 80 °C, 60 rpm and a 24 kg/h flow rate [27]. It would be worthwhile to directly compare the oil yields obtained using single versus twin screw extrusion as a pre- or co-treatment for expeller pressing. There is still more work to be conducted regarding correlating research results from lab-scale expeller pressing of various seeds and scaling up to industrial presses [17].

3. Solvent Extraction

Hexane (or *n*-hexane, which is its isomerized form) is the most commonly used organic solvent in the oilseed extraction industry due to its efficiency in oil recovery, inexpensive costs, recyclability, non-polar nature, low heat of vaporization and low boiling point (63–67 °C) [28,29]. Hexane extraction is especially utilized to produce soybean oil, which is the most consumed vegetable oil in the U.S. [30]. However, hexane is explosive, making it unsafe for workers in food-processing plants. In addition, it is both a neurological toxin and a hazardous air pollutant and can cause environmental pollution [31]. Although it is feasible to minimize these concerns with proper precautions, the production of certain foods, such as all organic foods, is restricted from hexane use. There is indeed evidence that hexane residue up to 21 ppm can be found in soy ingredients and in the 1 ppm range in vegetable oils [32,33]. For soy foods, the Food and Drug Administration (FDA) has not set maximum hexane residue limits, but the European Union (EU) prohibits hexane residue levels greater than 10 ppm. The EU has also set other hexane residue limits that vary depending on the food product [34].

There are many examples of the use of co-solvents during solvent extractions [35,36]; however, in this review we are only covering studies that discuss single solvent extractions. Alternatives to hexane such as ethanol (a natural, non-toxic solvent allowed in organic food production) have been investigated [37]. As ethanol is more polar than hexane, it has the capability to extract more polar compounds such as polyphenols, pigments and soluble sugars. The benefit of using ethanol vs. *n*-hexane was demonstrated during the extraction of sunflower collets (ground oilcake or expanded material), with a 32% vs. 23% yield of extracted material (oil and other compounds), respectively [38]. With sunflower collets, ethanol extraction led to a 38% greater extractability of tocopherols and phospholipids compared to *n*-hexane [38]. With castor seeds, no significant difference was found in the oil yield between hexane and ethanol, but the extract obtained with ethanol had significantly higher level of sterols than in the hexane extract [39].

Oil extraction yields can be improved using other organic solvents besides ethanol. Isopropanol extracted 49 wt.% oil from favela seeds, which was significantly higher than 47% using *n*-hexane [40]. The extraction method itself can have an important impact on the choice of the organic solvent leading to higher oil extraction yield. When ultrasound, shaker and Soxhlet methods utilizing hexane, acetone, ethanol and isopropanol were compared for the extraction of passion fruit oil, the highest oil yield (26%) was obtained using hexane during Soxhlet extraction. Acetone was the most effective solvent for ultrasound extraction (24% vs. 17% when using hexane) [15]. Ethyl acetate has characteristics that could be beneficial, as it is less flammable and hazardous and 33% cheaper compared to *n*-hexane. Similar oil extraction yields were obtained during the extraction of canola seeds with hexane (21–36%) and ethyl acetate (25 to 40%), while for camelina seeds, it ranged from 9–16% for both solvents [41]. Thus, the literature provides evidence that alternative organic solvents could replace the use of hexane for a similar oil extraction yield.

4. Supercritical CO_2 Extraction

There has been a rise of SCO_2 technology over the past few decades, with over 150 supercritical fluid extraction plants located around the world in 2014, mostly in North America and Europe [42]. SCO_2 is used to de-caffeinate coffee and tea and extract oils, antioxidants, natural food colorings, aromas and flavors from various food matrices [43]. For oil extraction in particular, it has been applied to a wide variety of seeds such as apricot, canola, soybean, sunflower, grape, acorn and walnut seeds [44]. During SCO_2 extraction, pressurized CO_2 solvent is mixed with solid raw material (often ground to reduce the particle size), which allows for the extraction of the compounds of interest. A pressurized CO_2 solvent begins to form at its critical point of 31 °C and 7.38 MPa, where the gas and liquid phases come together to form a homogeneous fluid phase beyond the supercritical fluid region. The advantages of SCO_2 extraction over conventional solvent extraction methods include higher diffusivity, lower viscosity and surface tension and faster extraction times [43]. Additionally, using CO_2 has environmental benefits such as being nonflammable and recyclable. It allows for improved product quality by leaving no residues and maintaining high purity of extracted materials. For these reasons, it is often considered a "greener" extraction method compared to solvent extraction [45].

However, there are several pitfalls to using SCO_2. The non-polarity of CO_2 limits extraction capabilities of polar phytochemicals, such as phenols [46,47]. This extraction method is also expensive because it relies on equipment that handles high pressure, which increases investment and maintenance costs [48]. Additionally, there is a lack and need for continuous systems for increasing large-scale production capacity. For these reasons, widespread adoption of this extraction method by the food industry has been lagging [43,49]. While having benefits that other technologies do not have, the economical competitivity of SCO_2 is a major pitfall for its development [42].

The selection of the most favorable SCO_2 extraction parameters for pressure, temperature, solvent flow rate, size of materials and moisture content are dependent on the type of seed and molecules of interest. For hemp seed oil, increasing the pressure reduced the extraction time (4.5 h at 30 MPa vs. 3.5 h at 40 MPa), while increasing the temperature from 40 to 60 °C did not significantly impact the extraction yield [50]. Oil extraction of peach seeds was improved by decreasing the particle size and increasing the temperature, flow rate, pressure and extraction time. Applying SCO_2 for 3 h at conditions of 40 °C, 20 MPa and 7 ml/min to 0.3 mm ground peach seeds led to a 35% oil yield. This was within the range reported for peach seeds extracted using solvents [51,52].

Sometimes the benefit of SCO_2 over solvents is observed in oil quality rather than yield. When extracting samara oil from different cultivars of *Elaeugnus mollis* Diels seeds using SCO_2, the oil yields ranged from 25–38%, which were significantly lower than with hexane (47–52%). However, the use of SCO_2 led to a higher quality oil due to the greater extraction of unsaturated fatty acids, such as linoleic acid, which promotes brain function [53]. Although the oil yield from Moringa seeds from petroleum ether extraction surpassed that of SCO_2 extractions, there was no significant difference in fatty acids, tocopherols and sterols between the two methods [54]. The SCO_2 extraction of different grape seeds successfully led to 3.4–4.8 mg/kg extraction of lycopene, a carotenoid that serves as an antioxidant and precursor to vitamin A. This represented an approximately 20% increase when compared to hexane [55,56].

However, the low polarity of CO_2 can cause difficulty in the extraction of polar lipids such as phospholipids and phenols, and could be a major drawback for this technology. This can be overcome by combining polar co-solvents with SCO_2, which improves solubility of the solute during extraction via dipole–dipole and hydrogen bond interactions [57,58]. SCO_2 extraction of camelina seeds using ethanol as a co-solvent improved the extraction of phospholipids and phenols, and thus increased total lipid yields (34% vs. 23% for pure SCO_2 extraction) [59]. The highest total phenol content from grape seeds was achieved through a sequential SCO_2 extraction, in which non-polar components were removed from grape seeds first, and then 15% mol ethanol was added to recover phenols from defatted grape seeds. Thus ethanol-assisted SCO_2 extraction may provide oils with better health benefits by extracting higher yields of specific compounds [60].

5. Aqueous Extraction Processing

The main benefit of aqueous extractions (AEP) for seeds is that water can be used as a more environmentally friendly solvent compared to organic solvents such as hexane. With solvent extractions, the oil from the seed substrate is dissolved into the solvent phase. The oil is then recovered by the evaporation of the organic solvent. With AEP, oil is typically partitioned into the following fractions: solid residue, protein-rich skim, lipid-rich cream and free oil (Figure 3). Therefore, additional steps are necessary to release free oil, such as demulsification from the cream. However, these steps could add significant costs to oil recovery; therefore, the most ideal extraction would be one that extracts the most free oil. Hence, extraction yield is not always the best indicator of recovered free oil.

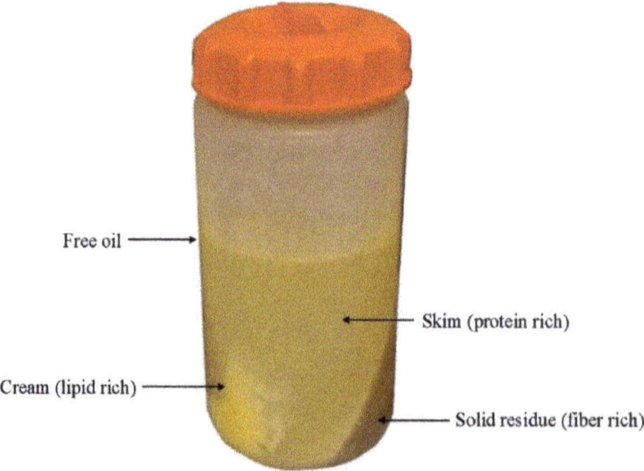

Figure 3. Image of partitioning in aqueous extraction slurry after centrifugation.

Regardless of the matrix, AEP usually has lower yields than the ones obtained with organic solvents; however, some research studies have shown competitive recovery yields up to 96% [61,62]. Pre-treatments are commonly applied prior to aqueous extractions, and all have the same goal of breaking down/softening the seed matrix to increase oil recovery. For example, roasting seeds can improve yields because applying heat to the substrate can rupture cell walls and allow for better oil release [63]. Thus, using the optimized roasting temperature and time of wild almonds led to an oil extraction yield of 35% (w/w) [61]. Flaking and extrusion prior to aqueous extraction can also promote increased cell disruption, allowing for better water penetration and release of entrapped compounds. For the aqueous extraction of soybeans, oil extraction yields improved significantly when using extruded full fat soybean flakes (68%) compared to non-treated soybean flakes (60%) [64]. Some pretreatments were shown to specifically improve free oil recovery. Flaxseed kernels pretreated with 0.3 M citric acid and dried at 70 °C for 1 h prior to aqueous extraction led to the development of a thinner cream layer and increased the free oil yields from 19 to 83%. This significant increase in free oil recovery was related to the ability of the acid treatment to affect protein properties, which led to the coalescence of oil bodies and size reduction in protein bodies [65]. Furthermore, different instruments can be used to perform aqueous oil extraction. The use of a twin-screw extruder for the aqueous extraction of sunflower seeds led to 35% higher oil yield than when processed in a blender [66].

Enzyme-assisted Aqueous Extraction (EAEP)

Enzymes are frequently used in all realms of food processing and their addition to aqueous oil extraction offers many advantages. Oil is difficult to release from the cotyledon, which is protected by cell wall structures made up of cellulose, hemicellulose, lignin and pectin [28]. Thus, seeds can be treated with substrate-specific enzymes such as carbohydrases (i.e., cellulase, hemicellulase and pectinase) to degrade the cell wall and facilitate oil release. Protease is utilized for hydrolyzing proteins in the cell membrane, which increases the extraction efficiency of seeds [67,68]. Enzyme treatment is environmentally friendly, occurs at mild temperatures and does not produce solvent residues [28,67,68]. Life cycle analyses (LCA) on enzyme usage in food, feed and pharmaceutical industries have demonstrated that enzymatic processes lead to less impact on global warming, acidification, eutrophication, ozone formation and energy consumption [69]. Enzymes can be expensive, but costs could be compensated by the increase in extraction yield or enzyme recycling [67,70].

As in any process, optimizing parameters is important to obtain good extraction yields. While the optimal pH of enzymes depends on many parameters including the type of enzymes (for example, proteases vs. carbohydrases), it is also crucial to set the pH far from the isoelectric pH (pI) of seed proteins. At their pI, proteins are insoluble, which can hinder oil extractions [28,71–73]. Temperature is another important parameter to consider when using enzymes, with the ideal range for enzymatic hydrolysis typically between 45–55 °C. If temperatures are too high, enzymes can become inactive and, thus, reduce their hydrolysis capabilities. However, temperatures that are too low can slow the reaction rates of enzymes and the extraction rate of oils [28,71,74,75].

The benefit of EAEP has been shown in comparison to AEP of seeds. For the oil extraction of Moringa seeds, individual addition of proteases and carbohydrases led to oil yields ranging from 17–23% compared to 8% for the control. Protease led to highest oil recovery due to its role in solubilizing proteins in the seed substrate [76]. Adding 0.85% alkaline protease to an almond cake slurry extracted for 1 h at pH 9 and 50 °C led to a significantly higher oil extraction (50%) compared to the non-enzymatic control extraction (42%) [77]. However, some studies show that enzyme addition does not improve oil yields [78]. For example, EAEP vs. AEP of almond cake extracted using the same enzymes and parameters mentioned above did not lead to significantly different oil extraction yields (26 and 29%, respectively). However, the change in scale may have contributed to the different outcomes; the first study was conducted at lab scale (50 g), while the second study was at pilot scale (750 g) [79].

Another approach is to use a cocktail of enzymes, added simultaneously, during the extraction. A cocktail of cellulase, pectinase and hemicellulase was used to extract oil from yellow mustard flour through a 3-h enzyme-assisted aqueous oil extraction set to pH 4.5–5.0 and 40–42 °C. Yields of 76% of oil and 75% of protein were reported, which were significantly higher compared to aqueous extraction leading to yields of 56% of oil and 61% of protein [73]. The oil extraction of *Forsythia suspense* seeds was improved using a cocktail of cellulase, pectinase and proteinase (17 vs. 7% for AEP, respectively). This improvement was attributed to more components within the seed cell walls being degraded [80].

Sometimes enzyme addition is more helpful during the demulsification step to release free oil from the cream layer. Due to the composition of the cream layer, proteases and phospholipases are often considered [81]. The mechanisms involved during enzymatic demulsification include hydrolysis of the proteins in the emulsion, leading to larger oil droplet coalescence and free oil recovery [71,73]. Enzymatic demulsification of cream using 0.5% alkaline protease after both AEP and EAEP of almond cake significantly improved the free oil yield (60–63%) compared to the control (up to 39%) [78]. Cream from peanut seed extraction was also destabilized using alkaline protease, achieving a 65% free oil yield. This was a steep increase compared to the cream from the control, which had less than a 5% free oil yield [82]. Additionally, enzymes can be used to increase protein recovery from the skim layer. Protease-based EAEP of almond flour set to pH 9 and then adjusted to pH 5 (the pI of almond proteins) led to the production of significantly more soluble peptides compared to AEP (45 vs. 23%, respectively) [79]. These examples highlight the potential for enzymes to be used on the various fractions of EAEP for multiple food applications.

6. Life Cycle and Environmental Impact Analyses

Each oil extraction method has its own advantages and disadvantages in terms of environmental, economic and practical aspects. Therefore, the best method depends on the matrix and specific desired outcomes. Life cycle analyses (LCA) is one way in which environmental impacts can be assessed for the lifetime of any given food product, from cradle-to-grave or even cradle-to-cradle. This tool has become more popular in the food sector in the last decade, which is evident by the increasing number of publishing frequency of LCA studies regarding food topics [83,84]. LCA, which is independent of time and location variables, is often contrasted to Environmental Impact Analysis (EIA). The latter is a tool that also considers the environmental impact of food products, but unlike LCA, it also covers social impacts, such as time-related or local geographic factors [85,86].

LCA studies on edible oils frequently cover oilseed crop cultivation, oil extraction and transportation within their system boundaries [87]. Yet, there is limited information on the environmental impacts between extraction methods. An LCA study on mustard seed oil demonstrated that extraction via pressing had significantly lower environmental impacts than a combination of pressing and solvent use. The latter method showed an 8–9% increase for several impact categories, such as human toxicity and particular matter potential. The impact based on photochemical oxidant formation potential increased by 15% due to hexane emissions [88]. When comparing the use of hexane vs. ethanol during the soybean oil extraction process, the net present value (the economic metric representing cash flow) was 10.2% higher for hexane extraction; however, the global warming potential for ethanol extraction was lower by 10,600 tons of CO_{2eq} per year [89].

Additionally, EIA has been performed to compare mass flows, energy consumption and global warming potential between hexane, expeller and EAEP methods for soybean oil processing (Table 2). As with mustard seeds, it was demonstrated that the use of hexane to extract soybean oil had the highest environmental impact. Additionally, hexane displayed a higher thermal risk and impacts on acute, chronic and eco-toxicity; however, hexane extraction was the lowest regarding air pollutant and greenhouse gas (GHG) emissions. The expelling process had the lowest environmental impact because it uses the least amount of chemical additives. However, its downside was generating the highest GHG (about 11 times more CO_2 and CH_4 emitted from 1 kg of soybean oil production compared to hexane extraction) and the highest criteria pollutant emissions due to the energy used during pressing. EAEP was concluded to be an ideal alternative candidate because it has lower environmental impacts compared to hexane extraction, and released less GHG and pollutants compared to expelling [90]. Although EAEP was again shown to lower environmental impacts in another study on soybean oil extraction, it had the highest CO_2 and GHG emissions compared to pressing and hexane, which was explained by the intensive electricity consumption used during the pretreatment (cleaning, drying, cracking, flaking and tempering) of the soybean substrate to maximize the oil yield. As a consequence, it was concluded that expelling, and not EAEP, was the cleanest oil extraction method [91].

Table 2. Comparison of published techno-economic analysis and environmental impact analysis studies on soybean oil extraction methods (adapted from [24,70,90,92,93]).

Extraction Method	Oil Yields	Profitable Capacity of Annual Oil Production	Revenue Sources	Environmental Impact Analysis
EAEP [1]	Over 80%	>17 million kg	24% from oil >70% from insoluble fibers	Similar environmental impacts to expelling Lower greenhouse gas and criteria pollutants emissions
Extruder-expelling	72%	13 million kg	23% from oil 77% from meal	Lowest environmental impacts Highest greenhouse gas and criteria pollutants emissions
Hexane	Over 99%	87 million kg	39% from oil >60% from meal and hulls	Highest environmental impact Most energy efficient Lowest greenhouse gas and criteria pollutants emissions

[1] Enzyme-assisted aqueous extraction processing.

7. Techno-Economic Analysis (TEA)

Economic feasibility is the main driver in the decision process on which extraction method to apply at an industrial scale; however, environmental impacts are increasingly being considered. Techno-economic analysis (TEA) allows the breakdown of profits and costs for any type of industrial process, and the analysis has been applied to oil extraction. A wide variety of parameters need to be considered when performing TEA, including, but not limited to, the scale of the extraction processes, type of substrate and extraction plant location [24]. Despite some variabilities in the outcomes of studies focusing on oil extraction methods, the valorization of co-products is of paramount importance for making profit and offsetting the costs for advanced extraction technology in processing plants [24,70,92]. When possible and economically viable, these co-products are widely utilized as animal feed and non-food applications instead of being discarded as food waste [24,92].

A good example illustrating this point is what occurs in the soybean oil industry (Table 2). During expeller pressing, the solid residue made of fiber and protein often has residual oil, which adds value to this co-product. When TEA was applied to a two-step extruder–expelling of soybean oil extraction, it was found that soybean meal was the driving force in profits, contributing 75% of total revenues [24]. Similarly, the importance of soybean co-products on the techno-economic value of EAEP extraction was demonstrated. Although EAEP led to an extraction yield that can compete with organic solvent extraction, the enzyme and facility costs for extraction and demulsification equipment were high. Soybean oil profits only accounted for 27% of total revenues, but co-product utilization in soybean/corn-based ethanol production made up 74% of total revenues. Other money saving practices may include recycling the enzymes and reusing the skim as a water source [70].

As with soybean oil extraction, the economic feasibility of alternative green extraction methods for other seeds were demonstrated to be dependent on their co-products. The values of the co-products (rapeseed meal and molasses) compensated for the higher crushing costs resulting from ethanol extraction compared to the hexane extraction of rapeseed oil [37]. The economics involved in the industrial scale of SCO_2 oil extraction of 3000 ton/yr of grape marc were calculated. Selling the dried skins and exhausted seed powder by-products for cattle feed garnered extra revenue of about 60 EUR/ton (USD 78/ton). The 2100 and 2700 tons/year of dried skins and exhausted seed powder produced, respectively, helped meet the breakeven point of 5.9 EUR/kg (USD 8/kg) [94]. In conclusion, the extraction yield and cost were not the sole indicators of the economic viability of an oil extraction process. Co-products were an important piece of the puzzle.

TEA is a powerful tool for evaluating facility scale-up and subsequent economics of oil extraction processes. For example, a TEA model on soybean oil hexane extraction identified that a plant capacity of 34.6 million kg of annual soybean oil production was needed for the process to financially breakeven [92]. For the EAEP of soybean oil to be profitable, annual oil production could not fall below 8.5 million kg [70]. SCO_2 extraction using two extractors in series increased production efficiency of grapeseed oil from 83–86% compared to one with only a larger single extractor. Food waste was also reduced by utilizing grape stalks and skin by-products as thermal energy during SCO_2 extraction. The grapeseed oil market value was reported to be as high as 30 EUR/kg (USD 39/kg). When making the assumption that SCO_2 extracted oil will have similar quality and thus comparable market value, it was demonstrated the oil has market potential [94]. Another simulation modeled the economics of vegetable oil extraction in a SCO_2 industrial plant investigating how adjusting the extraction time and the particle diameter of the substrate can alter costs. Maintaining a 2.3-h SCO_2 extraction time reduced the production cost to USD 9.4/kg oil. Increasing the particle size from 0.5 to 4 mm can decrease the extraction time from 5 to 3.6 h [42].

TEA is, therefore, an integral part of determining whether an oil extraction method is feasible at a commercial scale. Therefore, processors must consider substrate preparation and extraction flow processes to make large-scale, green oil extraction more economical [40]. However, it is important to note that there are limitations to studying industrial plant economics using a simulation approach. Models typically assume laboratory scale conditions, which set ideal parameters for substrates that are less feasible at commercial scale. Therefore, scale-up predictions must be validated, and more research is needed to refine the accuracy of cost estimates [42]. Further investigation and collaboration with industrial adopters of SCO_2 technology is required to better understand these large-scale oil extraction projects.

While TEA is crucial for identifying economic feasibility, more studies are now integrating an environmental component to it. A techno-economic study comparing subcritical water, SCO_2 and solvent extractions of bioactive compounds from grape marc emphasized the energy-intensive and costly aspect of SCO_2 extraction. SCO_2 extractions had the highest cost of manufacturing (USD 88/kg product) and the lowest net present value (-USD 920,000). SCO_2 extraction also had the highest environmental impact due to energy use (11.8 kg CO_2-eq/kg product), which countered the common perception that SCO_2 technology is more environmentally friendly [47]. However, comparative studies are lacking in the scientific literature, and future development of the edible oil industry would benefit from more techno-economic analyses between various extraction methods.

8. Conclusions

The food industry has increasingly promoted a circular, green economy by prioritizing sustainability and a reduction in food waste. One way to reduce waste is through the development of functional by-products. Specialty seeds have been a target for conversion to edible oils for human consumption. Several factors such as profitability and environmental sustainability should be addressed when determining which of the many existing extraction methods to implement. Ethanol and SCO_2 extractions are considered viable alternatives to using hexane. Additionally, the use of enzymes during the aqueous extraction of seeds allows for a process with less environmental risks compared to traditional hexane extraction. The implementation of economically feasible, greener extraction practices in an industry setting requires the valorization of co-products and optimization of extraction parameters. With increasing interest by consumers for sustainable food products, the specialty oil industry would benefit from improved extraction methods supported by both techno-economic studies and environmental impact and life cycle analyses.

Author Contributions: Conceptualization, S.J.; investigation, V.M.L. and S.J.; writing–original draft preparation, V.M.L., K.A.R. and S.J.; supervision, S.J., project and administration, S.J., funding acquisition, S.J. All authors have read and agreed to the published version of the manuscript.

Funding: This work was funded by the National Institute of Food and Agriculture, National Needs Graduate Fellowship Program, United States Department of Agriculture, No. 2017-38420-26767. Partial funding for this project has been made available by the California State University Agricultural Research Institute (ARI) project 19-03-120.

Institutional Review Board Statement: Not applicable.

Informed Consent Statement: Not applicable.

Data Availability Statement: Not applicable.

Conflicts of Interest: The authors declare no conflict of interest.

References

1. United States 2030 Food Loss and Waste Reduction Goal. Available online: https://www.epa.gov/sustainable-management-food/united-states-2030-food-loss-and-waste-reduction-goal (accessed on 23 September 2021).
2. Food Wastage: Key Facts and Figures. Available online: http://www.fao.org/news/story/en/item/196402/icode/ (accessed on 23 September 2021).
3. Food Recovery Hierarchy. Available online: https://www.epa.gov/sustainable-management-food/food-recovery-hierarchy (accessed on 23 September 2021).
4. Food Recovery Act. Available online: https://policyfinder.refed.org/federal-policy/food-recovery-act (accessed on 23 September 2021).
5. Short-Lived Climate Pollutants (SLCP): Organic Waste Methane Emissions Reductions. Available online: https://www.calrecycle.ca.gov/climate/slcp (accessed on 15 June 2021).
6. Food Loss and Food Waste. Available online: http://www.fao.org/3/i4807e/i4807e.pdf (accessed on 28 April 2020).
7. Gustavsson, J.; Cederberg, C.; Sonesson, U.; Emanuelsson, A. *The Methodology of the FAO Study: "Global Food Losses and Food Waste-Extent, Causes and Prevention"*; FAO: Sweden, 2013. Available online: http://www.diva-portal.org/smash/get/diva2:944159/FULLTEXT01.pdf (accessed on 18 July 2021).
8. Oilseeds: World Market and Trade. Available online: https://downloads.usda.library.cornell.edu/usda-esmis/files/tx31qh68h/8g84nh663/3r075q71r/oilseeds.pdf (accessed on 25 September 2021).
9. Edible Oil & Fats Market Worth $162.51 Billion by 2025 | CAGR 7.6%. Available online: https://www.grandviewresearch.com/press-release/global-edible-oil-fats-market (accessed on 15 June 2021).
10. Turrini, F.; Zunin, P.; Boggia, R. Potentialities of rapid analytical strategies for the identification of the botanical species of several specialty or gourmet oils. *Foods* **2021**, *10*, 183. [CrossRef]
11. Gulcin, İ. Antioxidants and antioxidant methods: An updated overview. *Arch. Toxicol.* **2020**, *94*, 651–715. [CrossRef] [PubMed]
12. Global Vegetable Oil Consumption, 2019/20. Available online: https://www.statista.com/statistics/263937/vegetable-oils-global-consumption/ (accessed on 16 June 2021).
13. Gourmet and Specialty Oils. Available online: https://extension.okstate.edu/fact-sheets/gourmet-and-specialty-oils.html (accessed on 16 June 2021).
14. Cheng, M.H.; Dien, B.S.; Singh, V. Economics of plant oil recovery: A review. *Biocatal. Agric. Biotechnol.* **2019**, *18*, 101056. [CrossRef]
15. De Oliveira, R.C.; Davantel De Barros, S.T.; Gimenes, M.L. The extraction of passion fruit oil with green solvents. *J. Food Eng.* **2013**, *117*, 458–463. [CrossRef]
16. Kasote, D.M.; Badhe, Y.S.; Hegde, M.V. Effect of mechanical press oil extraction processing on quality of linseed oil. *Ind. Crops Prod.* **2013**, *42*, 10–13. [CrossRef]
17. Savoire, R.; Lanoisellé, J.L.; Vorobiev, E. Mechanical continuous oil expression from oilseeds: A review. *Food Bioprocess Technol.* **2013**, *6*, 1–16. [CrossRef]
18. Oilseed Fact Sheet: Oilseed Presses. Available online: https://www.uvm.edu/sites/default/files/media/Oilseed_Presses.pdf (accessed on 23 September 2021).

19. Expanding and Expelling. Available online: https://lipidlibrary.aocs.org/edible-oil-processing/expanding-and-expelling (accessed on 21 August 2021).
20. Bogaert, L.; Mathieu, H.; Mhemdi, H.; Vorobiev, E. Characterization of oilseeds mechanical expression in an instrumented pilot screw press. *Ind. Crops Prod.* **2018**, *121*, 106–113. [CrossRef]
21. Romuli, S.; Karaj, S.; Latif, S.; Müller, J. Performance of mechanical co-extraction of *Jatropha curcas* L. kernels with rapeseed, maize or soybean with regard to oil recovery, press capacity and product quality. *Ind. Crops Prod.* **2017**, *104*, 81–90. [CrossRef]
22. Uitterhaegen, E.; Evon, P. Twin-screw extrusion technology for vegetable oil extraction: A review. *J. Food Eng.* **2017**, *212*, 190–200. [CrossRef]
23. Vandenbossche, V.; Candy, L.; Evon, P.; Rouilly, A.; Pontalier, P.Y. Extrusion. In *Green Food Processing Techniques: Preservation, Transformation, and Extraction*; Elsevier: Amsterdam, The Netherlands, 2019; pp. 289–314.
24. Cheng, M.H.; Rosentrater, K.A. Techno-economic analysis of extruding-expelling of soybeans to produce oil and meal. *Agriculture* **2019**, *9*, 87. [CrossRef]
25. Amalia Kartika, I.; Pontalier, P.Y.; Rigal, L. Twin-screw extruder for oil processing of sunflower seeds: Thermo-mechanical pressing and solvent extraction in a single step. *Ind. Crops Prod.* **2010**, *32*, 297–304. [CrossRef]
26. Sriti, J.; Msaada, K.; Talou, T.; Faye, M.; Kartika, I.A.; Marzouk, B. Extraction of coriander oil by twin-screw extruder: Screw configuration and operating conditions effect. *Ind. Crops Prod.* **2012**, *40*, 355–360. [CrossRef]
27. Kartika, I.A.; Pontalier, P.Y.; Rigal, L. Oil extraction of oleic sunflower seeds by twin screw extruder: Influence of screw configuration and operating conditions. *Ind. Crops Prod.* **2005**, *22*, 207–222. [CrossRef]
28. Kumar, S.P.J.; Prasad, S.R.; Banerjee, R.; Agarwal, D.K.; Kulkarni, K.S.; Ramesh, K.V. Green solvents and technologies for oil extraction from oilseeds. *Chem. Cent. J.* **2017**, *11*, 1–7. [CrossRef]
29. Li, Y.; Fine, F.; Fabiano-Tixier, A.S.; Abert-Vian, M.; Carre, P.; Pages, X.; Chemat, F. Evaluation of alternative solvents for improvement of oil extraction from rapeseeds. *Comptes Rendus Chim.* **2014**, *17*, 242–251. [CrossRef]
30. Soystats. Available online: https://soygrowers.com/wp-content/uploads/2019/10/Soy-Stats-2019_FNL-Web.pdf (accessed on 22 September 2021).
31. Hexane. Available online: https://www.epa.gov/sites/production/files/2016-09/documents/hexane.pdf (accessed on 21 June 2021).
32. Toxic Chemicals: Banned in Organics But Common in "Natural" Food Production. Available online: https://www.cornucopia.org/hexane-guides/nvo_hexane_report.pdf (accessed on 23 September 2021).
33. Yousefi, M.; Yousefi, M.; Hosseini, H. Evaluation of hexane content in edible vegetable oils consumed in Iran. *J. Exp. Clin. Toxicol.* **2017**, *1*, 27–30. [CrossRef]
34. Directive 2009/32/EC of the European Parliament and of the Council of 23 April 2009 on the Approximation of the Laws of the Member States on Extraction Solvents Used in the Production of Foodstuffs and Food Ingredients. Available online: https://webarchive.nationalarchives.gov.uk/eu-exit/https://eur-lex.europa.eu/legal-content/EN/TXT/?uri=CELEX:02009L0032-20161109 (accessed on 23 September 2021).
35. Dagostin, J.L.A.; Carpiné, D.; Corazza, M.L. Extraction of soybean oil using ethanol and mixtures with alkyl esters (biodiesel) as co-solvent: Kinetics and thermodynamics. *Ind. Crops Prod.* **2015**, *74*, 69–75. [CrossRef]
36. Jin, C.; Zhang, X.; Geng, Z.; Pang, X.; Wang, X.; Ji, J.; Wang, G.; Liu, H. Effects of various co-solvents on the solubility between blends of soybean oil with either methanol or ethanol. *Fuel* **2019**, *244*, 461–471. [CrossRef]
37. Carré, P.; Citeau, M.; Dauguet, S. Hot ethanol extraction: Economic feasibility of a new and green process. *OCL* **2018**, *25*, 1–7. [CrossRef]
38. Baümler, E.R.; Carrín, M.E.; Carelli, A.A. Extraction of sunflower oil using ethanol as solvent. *J. Food Eng.* **2016**, *178*, 190–197. [CrossRef]
39. Sbihi, H.M.; Nehdi, I.A.; Mokbli, S.; Romdhani-Younes, M.; Al-Resayes, S.I. Hexane and ethanol extracted seed oils and leaf essential compositions from two castor plant (*Ricinus communis* L.) varieties. *Ind. Crops Prod.* **2018**, *122*, 174–181. [CrossRef]
40. Santos, K.A.; da Silva, E.A.; da Silva, C. Ultrasound-assisted extraction of favela (*Cnidoscolus quercifolius*) seed oil using ethanol as a solvent. *J. Food Process. Preserv.* **2021**, *46*, e15497.
41. Lohani, U.C.; Fallahi, P.; Muthukumarappan, K. Comparison of ethyl acetate with hexane for oil extraction from various oilseeds. *J. Am. Oil Chem. Soc.* **2015**, *92*, 743–754. [CrossRef]
42. Del Valle, J.M. Extraction of natural compounds using supercritical CO_2: Going from the laboratory to the industrial application. *J. Supercrit. Fluids* **2015**, *96*, 180–199. [CrossRef]
43. Ye, X.; Xue, S.J.; Shi, J. Green separation technology in food processing: Supercritical-CO_2 fluid extraction. In *Advances in Food Processing Technology*; Jia, J., Liu, D., Ma, H., Eds.; Springer: Singapore, 2019; pp. 73–99.
44. Sánchez-Vicente, Y.; Cabañas, A.; Renuncio, J.A.R.; Pando, C. Supercritical fluid extraction of peach (*Prunus persica*) seed oil using carbon dioxide and ethanol. *J. Supercrit. Fluids* **2009**, *49*, 167–173. [CrossRef]
45. Smigic, N.; Djekic, I.; Tomic, N.; Udovicki, B.; Rajkovic, A. The potential of foods treated with supercritical carbon dioxide (SCO2) as novel foods. *Br. Food J.* **2019**, *121*, 815–834. [CrossRef]
46. Mushtaq, M.; Sultana, B.; Anwar, F.; Adnan, A.; Rizvi, S.S.H. Enzyme-assisted supercritical fluid extraction of phenolic antioxidants from pomegranate peel. *J. Supercrit. Fluids* **2015**, *104*, 122–131. [CrossRef]
47. Todd, R.; Baroutian, S. A techno-economic comparison of subcritical water, supercritical CO_2 and organic solvent extraction of bioactives from grape marc. *J. Clean. Prod.* **2017**, *158*, 349–358. [CrossRef]

58. Carvalho, P.I.N.; Osorio-Tobón, J.F.; Rostagno, M.A.; Petenate, A.J.; Meireles, M.A.A. Techno-economic evaluation of the extraction of turmeric (*Curcuma longa* L.) oil and ar-turmerone using supercritical carbon dioxide. *J. Supercrit. Fluids* **2015**, *105*, 44–54. [CrossRef]

59. İncedayi, B.; Suna, S.; Çopur, Ö.U. Use of supercritical CO$_2$ in food industry. *Bulg. Chem. Commun.* **2014**, *46*, 126–130.

60. Aladić, K.; Jarni, K.; Barbir, T.; Vidović, S.; Vladić, J.; Bilić, M.; Jokić, S. Supercritical CO$_2$ extraction of hemp (*Cannabis sativa* L.) seed oil. *Ind. Crops Prod.* **2015**, *76*, 472–478. [CrossRef]

61. Ekinci, M.S.; Gürü, M. Extraction of oil and β-sitosterol from peach (*Prunus persica*) seeds using supercritical carbon dioxide. *J. Supercrit. Fluids* **2014**, *92*, 319–323. [CrossRef]

62. Wu, H.; Shi, J.; Xue, S.; Kakuda, Y.; Wang, D.; Jiang, Y.; Ye, X.; Li, Y.; Subramanian, J. Essential oil extracted from peach (*Prunus persica*) kernel and its physicochemical and antioxidant properties. *LWT-Food Sci. Technol.* **2011**, *44*, 2032–2039. [CrossRef]

63. Mu, J.; Wu, G.; Chen, Z.; Brennan, C.S.; Tran, K.; Dilrukshi, H.N.N.; Shi, C.; Zhen, H.; Hui, X. Identification of the fatty acids profiles in supercritical CO$_2$ fluid and soxhlet extraction of samara oil from different cultivars of *Elaeagnus mollis* Diels seeds. *J. Food Compos. Anal.* **2021**, *101*, 1–9. [CrossRef]

64. Ruttarattanamongkol, K.; Siebenhandl-Ehn, S.; Schreiner, M.; Petrasch, A.M. Pilot-scale supercritical carbon dioxide extraction, physico-chemical properties and profile characterization of *Moringa oleifera* seed oil in comparison with conventional extraction methods. *Ind. Crops Prod.* **2014**, *58*, 68–77. [CrossRef]

65. Caseiro, M.; Ascenso, A.; Costa, A.; Creagh-Flynn, J.; Johnson, M.; Simões, S. Lycopene in human health. *LWT* **2020**, *127*, 109323. [CrossRef]

66. Mohamed, H.B.; Duba, K.S.; Fiori, L.; Abdelgawed, H.; Tlili, I.; Tounekti, T.; Zrig, A. Bioactive compounds and antioxidant activities of different grape (*Vitis vinifera* L.) seed oils extracted by supercritical CO$_2$ and organic solvent. *LWT* **2016**, *74*, 557–562. [CrossRef]

67. Güçlü-Üstündağ, Ö.; Temelli, F. Solubility behavior of ternary systems of lipids, cosolvents and supercritical carbon dioxide and processing aspects. *J. Supercrit. Fluids* **2005**, *36*, 1–15. [CrossRef]

68. Asep, E.K.; Jinap, S.; Jahurul, M.H.A.; Zaidul, I.S.M.; Singh, H. Effects of polar cosolvents on cocoa butter extraction using supercritical carbon dioxide. *Innov. Food Sci. Emerg. Technol.* **2013**, *20*, 152–160. [CrossRef]

69. Belayneh, H.D.; Wehling, R.L.; Reddy, A.K.; Cahoon, E.B.; Ciftci, O.N. Ethanol-modified supercritical carbon dioxide extraction of the bioactive lipid components of *Camelina sativa* seed. *J. Am. Oil Chem. Soc.* **2017**, *94*, 855–865. [CrossRef]

70. Da Porto, C.; Natolino, A. Supercritical fluid extraction of polyphenols from grape seed (*Vitis vinifera*): Study on process variables and kinetics. *J. Supercrit. Fluids* **2017**, *130*, 239–245. [CrossRef]

71. Moghadas, H.C.; Rezaei, K. Laboratory-scale optimization of roasting conditions followed by aqueous extraction of oil from wild almond. *J. Am. Oil Chem. Soc.* **2017**, *94*, 867–876. [CrossRef]

72. Fu, S.; Wu, W. Optimization of conditions for producing high-quality oil and de-oiled meal from almond seeds by water. *J. Food Process. Preserv.* **2019**, *43*, e14050. [CrossRef]

73. Akinoso, R.; Raji, A.O. Optimization of oil extraction from locust bean using response surface methodology. *Eur. J. Lipid Sci. Technol.* **2011**, *113*, 245–252. [CrossRef]

74. Jung, S.; Mahfuz, A.A. Low temperature dry extrusion and high-pressure processing prior to enzyme-assisted aqueous extraction of full fat soybean flakes. *Food Chem.* **2009**, *114*, 947–954. [CrossRef]

75. Zhang, W.; Peng, H.; Sun, H.; Hua, X.; Zhao, W.; Yang, R. Effect of acidic moisture-conditioning as pretreatment for aqueous extraction of flaxseed oil with lower water consumption. *Food Bioprod. Process.* **2020**, *121*, 20–28. [CrossRef]

76. Evon, P.; Vandenbossche, V.; Pontalier, P.Y.; Rigal, L. Direct extraction of oil from sunflower seeds by twin-screw extruder according to an aqueous extraction process: Feasibility study and influence of operating conditions. *Ind. Crops Prod.* **2007**, *26*, 351–359. [CrossRef]

77. Liu, J.J.; Gasmalla, M.A.A.; Li, P.; Yang, R. Enzyme-assisted extraction processing from oilseeds: Principle, processing and application. *Innov. Food Sci. Emerg. Technol.* **2016**, *35*, 184–193. [CrossRef]

78. Valladares-Diestra, K.; de Souza Vandenberghe, L.P.; Soccol, C.R. Oilseed enzymatic pretreatment for efficient oil recovery in biodiesel production industry: A review. *Bioenergy Res.* **2020**, *13*, 1016–1030. [CrossRef]

79. Jegannathan, K.R.; Nielsen, P.H. Environmental assessment of enzyme use in industrial production—A literature review. *J. Clean. Prod.* **2013**, *42*, 228–240. [CrossRef]

80. Cheng, M.H.; Rosentrater, K.A.; Sekhon, J.; Wang, T.; Jung, S.; Johnson, L.A. Economic feasibility of soybean oil production by enzyme-assisted aqueous extraction processing. *Food Bioprocess Technol.* **2019**, *12*, 539–550. [CrossRef]

81. Yusoff, M.; Gordon, M.H.; Niranjan, K. Aqueous enzyme assisted oil extraction from oilseeds and emulsion de-emulsifying methods: A review. *Trends Food Sci. Technol.* **2015**, *41*, 60–82. [CrossRef]

82. Mwaurah, P.W.; Kumar, S.; Kumar, N.; Kumar, A.; Panghal, A.A.; Kumar, V.; Mukesh, S.; Garg, K. Novel oil extraction technologies: Process conditions, quality parameters, and optimization. *Compr. Rev. Food Sci. Food Saf.* **2019**, *19*, 3–20. [CrossRef]

83. Tabtabaei, S.; Diosady, L.L. Aqueous and enzymatic extraction processes for the production of food-grade proteins and industrial oil from dehulled yellow mustard flour. *Food Res. Int.* **2013**, *52*, 547–556. [CrossRef]

84. Zúñiga, M.E.; Soto, C.; Mora, A.; Chamy, R.; Lema, J.M. Enzymic pre-treatment of *Guevina avellana* mol oil extraction by pressing. *Process Biochem.* **2003**, *39*, 51–57. [CrossRef]

75. Rui, H.; Zhang, L.; Li, Z.; Pan, Y. Extraction and characteristics of seed kernel oil from white pitaya. *J. Food Eng.* **2009**, *93*, 482–486. [CrossRef]

76. Latif, S.; Anwar, F.; Hussain, A.I.; Shahid, M. Aqueous enzymatic process for oil and protein extraction from *Moringa oleifera* seed. *Eur. J. Lipid Sci. Technol.* **2011**, *113*, 1012–1018. [CrossRef]

77. Souza, T.S.P.; Dias, F.F.G.; Koblitz, M.G.B.; Juliana, J.M.L.N. Aqueous and enzymatic extraction of oil and protein from almond cake: A comparative study. *Processes* **2019**, *7*, 472. [CrossRef]

78. De Almeida, N.M.; Dias, F.F.G.; Rodrigues, M.I.; de Moura Bell, J.M.L.N. Effects of processing conditions on the simultaneous extraction and distribution of oil and protein from almond flour. *Processes.* **2019**, *7*, 844. [CrossRef]

79. de Souza, T.S.P.; Dias, F.F.G.; Koblitz, M.G.B.; de Moura Bell, J.M.L.N. Effects of enzymatic extraction of oil and protein from almond cake on the physicochemical and functional properties of protein extracts. *Food Bioprod. Process.* **2020**, *122*, 280–290. [CrossRef]

80. Gai, Q.Y.; Jiao, J.; Wei, F.Y.; Luo, M.; Wang, W.; Zu, Y.G.; Fu, Y.J. Enzyme-assisted aqueous extraction of oil from *Forsythia suspensa* seed and its physicochemical property and antioxidant activity. *Ind. Crops Prod.* **2013**, *51*, 274–278. [CrossRef]

81. Yao, L.; Jung, S. 31P NMR Phospholipid profiling of soybean emulsion recovered from aqueous extraction. *J. Agric. Food Chem.* **2010**, *58*, 4866–4872. [CrossRef]

82. Zhang, S.B.; Liu, X.J.; Lu, Q.Y.; Wang, Z.W.; Zhao, X. Enzymatic demulsification of the oil-rich emulsion obtained by aqueous extraction of peanut seeds. *J. Am. Oil Chem. Soc.* **2013**, *90*, 1261–1270. [CrossRef]

83. Vidergar, P.; Perc, M.; Lukman, R.K. A survey of the life cycle assessment of food supply chains. *J. Clean. Prod.* **2021**, *286*, 125506. [CrossRef]

84. Muralikrishna, I.V.; Manickam, V. Life cycle assessment. *Environ. Manage* **2017**, 57–75.

85. Jeswani, H.K.; Azapagic, A.; Schepelmann, P.; Ritthoff, M. Options for broadening and deepening the LCA approaches. *J. Clean. Prod.* **2010**, *18*, 120–127. [CrossRef]

86. Morero, B.; Rodriguez, M.B.; Campanella, E.A. Environmental impact assessment as a complement of life cycle assessment. Case study: Upgrading of biogas. *Bioresour. Technol.* **2015**, *190*, 402–407. [CrossRef]

87. Khatri, P.; Jain, S. Environmental life cycle assessment of edible oils: A review of current knowledge and future research challenges. *J. Clean. Prod.* **2017**, *152*, 63–76. [CrossRef]

88. Khatri, P.; Jain, S.; Pandey, S. A cradle-to-gate assessment of environmental impacts for production of mustard oil using life cycle assessment approach. *J. Clean. Prod.* **2017**, *166*, 988–997. [CrossRef]

89. Potrich, E.; Miyoshi, S.C.; Machado, P.F.S.; Furlan, F.F.; Ribeiro, M.P.A.; Tardioli, P.W.; Giordano, R.L.C.; Cruz, A.J.G.; Giordano, R.C. Replacing hexane by ethanol for soybean oil extraction: Modeling, simulation, and techno-economic-environmental analysis. *J. Clean. Prod.* **2020**, *244*, 118660. [CrossRef]

90. Cheng, M.H.; Sekhon, J.J.K.; Rosentrater, K.A.; Wang, T.; Jung, S.; Johnson, L.A. Environmental impact assessment of soybean oil production: Extruding-expelling process, hexane extraction and aqueous extraction. *Food Bioprod. Process.* **2018**, *108*, 58–68. [CrossRef]

91. Cheng, M.H.; Zhang, W.; Rosentrater, K.; Sekhon, J.; Wang, T.; Jung, S.; Johnson, L. Environmental impact analysis of soybean oil production from expelling, hexane extraction and enzyme assisted aqueous extraction. In Proceedings of the 2016 ASABE Annual International Meeting, Orlando, FL, USA, 17–20 July 2016; p. 484.

92. Cheng, M.H.; Rosentrater, K.A. Economic feasibility analysis of soybean oil production by hexane extraction. *Ind. Crops Prod.* **2017**, *108*, 775–785. [CrossRef]

93. Cheng, M.H.; Rosentrater, K.A. Profitability analysis of soybean oil processes. *Bioengineering* **2017**, *4*, 83. [CrossRef]

94. Fiori, L. Supercritical extraction of grape seed oil at industrial-scale: Plant and process design, modeling, economic feasibility. *Chem. Eng. Process. Process Intensif.* **2010**, *49*, 866–872. [CrossRef]

MDPI

St. Alban-Anlage 66

4052 Basel

Switzerland

Tel. +41 61 683 77 34

Fax +41 61 302 89 18

www.mdpi.com

Processes Editorial Office

E-mail: processes@mdpi.com

www.mdpi.com/journal/processes